DAC Guidelines and Reference Series

W0230242

Environmental Fiscal Reform for Poverty Reduction

OECD

ORGANISATION FOR ECONOMIC CO-OPERATION AND DEVELOPMENT

ORGANISATION FOR ECONOMIC CO-OPERATION AND DEVELOPMENT

The OECD is a unique forum where the governments of 30 democracies work together to address the economic, social and environmental challenges of globalisation. The OECD is also at the forefront of efforts to understand and to help governments respond to new developments and concerns, such as corporate governance, the information economy and the challenges of an ageing population. The Organisation provides a setting where governments can compare policy experiences, seek answers to common problems, identify good practice and work to co-ordinate domestic and international policies.

The OECD member countries are: Australia, Austria, Belgium, Canada, the Czech Republic, Denmark, Finland, France, Germany, Greece, Hungary, Iceland, Ireland, Italy, Japan, Korea, Luxembourg, Mexico, the Netherlands, New Zealand, Norway, Poland, Portugal, the Slovak Republic, Spain, Sweden, Switzerland, Turkey, the United Kingdom and the United States. The Commission of the European Communities takes part in the work of the OECD.

OECD Publishing disseminates widely the results of the Organisation's statistics gathering and research on economic, social and environmental issues, as well as the conventions, guidelines and standards agreed by its members.

Also available in French under the title:

Lignes directrices et ouvrages de référence du CAD

LA RÉFORME FISCALE ÉCOLOGIQUE AXÉE SUR LA RÉDUCTION DE LA PAUVRETÉ

Foreword

*I*t is an unfortunate fact that fiscal and environmental experts seldom communicate with one another. They each have very different concerns, formal training and sets of jargon. You could even say that they speak a different language. But so much is to be gained if we could create incentives for them to work more together. Examples abound of thoughtfully designed fiscal measures (so-called market-based instruments) that, when effectively implemented, can raise revenue while helping to further environmental goals. This paper tells of many examples, and offers much new material highlighting the potential poverty reduction benefits – surely an added incentive to work together.

If joint reform of fiscal and environmental measures – or "environmental fiscal reform" (EFR) – can protect the environment, as well as raise revenue and free up resources – which can be allocated to poverty reduction efforts – why don't we see much more of it? For a start, designing effective EFR requires a sound understanding of environmental and fiscal policy, as well as how each is regulated in practice. This requires co-operation between fiscal and environmental experts, as well as colleagues in development. But designing a sound proposal for EFR is only half the story. Experience shows that it is also necessary to work across – and often to negotiate – hard political, social and institutional barriers if proposals are to be effectively implemented.

The OECD has proven to be well-placed in mobilising fiscal, environmental, and development experts to help us identify approaches to EFR that will work well in most developing countries. The experts have paid particular attention to the "political economy" of EFR. This includes identifying likely "winners" and "losers" from proposed reforms, how coalitions for change can be developed and nurtured, and how best to deal with those likely to oppose progress. The role of donors in supporting the process of EFR is also highlighted.

Preparing this document has been a challenge because it is rooted, not in tidy textbook solutions, but in "real world" experiences. I think it has been well worth the effort. I hope you will agree. More importantly, I hope that the analysis and recommendations contained in the document will be of use, whether you are a policy maker, a representative of a civil society group or the private sector in developing country partners. Sustainable development is about achieving integrated win-wins between environmental, economic and social objectives – but too often it has to handle trade-offs where such possibilities elude us. With real application of the guidance in this paper, I suspect we will see many more win-win cases emerge in the not too distant future.

Steve Bass
Chair, ENVIRONET

In order to achieve its aims the OECD has set up a number of specialised committees. One of these is the **Development Assistance Committee**, whose members have agreed to secure an expansion of aggregate volume of resources made available to developing countries and to improve their effectiveness. To this end, members periodically review together both the amount and the nature of their contributions to aid programmes, bilateral and multilateral, and consult each other on all other relevant aspects of their development assistance policies.

The members of the Development Assistance Committee are Australia, Austria, Belgium, Canada, Denmark, Finland, France, Germany, Greece, Ireland, Italy, Japan, Luxembourg, the Netherlands, New Zealand, Norway, Portugal, Spain, Sweden, Switzerland, the United Kingdom, the United States and the Commission of the European Communities.

ENVIRONMENTAL FISCAL REFORM FOR POVERTY REDUCTION – ISBN 92-64-008683 – © OECD 2005

Acknowledgements

T his document is the result of work undertaken by the OECD/DAC Network on Development Co-operation and Environment (ENVIRONET).

The document was prepared by a team led by the United Kingdom (Mr. Paul Steele and Mr. Richard Boyd) and Germany (Mr. Stephan Paulus, Mr. Harald Lossack and Mr. Jan-Peter Schemmel), supported by the Netherlands (Mr. Piet Klop), Sweden (Mr. Thomas Sterner, Göteborg University), the International Monetary Fund (Mr. Jim Prust), and the World Bank (Mr. Jan Bojö). Experts from China (Mr. Zhong Ma, Renmin University of China), India (Ms. Divya Datt, Tata Energy Research Institute; Mr. Subhash Garg, Ministry of Finance and Company Affairs; Mr. K.K. Narang, Ministry of Environment and Forests) and South Africa (Mr. Cecil Morden, National Treasury) provided valuable input. The OECD directorates responsible for Environment (Mr. Nils Axel Braathen), Agriculture and Fisheries (Mr. Carl-Christian Schmidt and Mr. Bertrand Le Gallic), Fiscal Affairs (Mr. Chris Heady) and Trade (Mr. Ron Steenblik), the International Energy Agency (Ms. Kristi Varangu) and several other donor agencies have also participated actively throughout the whole process. Messrs. Remy Paris and Georg Caspary of the OECD Secretariat provided substantive and managerial assistance to the expert group, while Maria Consolati provided invaluable secretarial assistance. Mr. Steve Bass, as Chair of ENVIRONET, provided guidance and advice throughout the exercise.

Four workshops were crucial in the development of this document:

- Scoping Workshop on Environmental Fiscal Reform (held on 30-31 January 2003 in Paris, and hosted by the OECD).
- Fiscal Reform for Fisheries: To Promote Growth, Poverty Eradication and Sustainable Development (held on 13-15 October 2003 in Rome, and hosted by the Food and Agriculture Organization of the UN-FAO).
- Reforming Forest Fiscal Systems to Promote Poverty Reduction and Sustainable Development (held on 19-21 October 2003 in Washington, DC, USA). This workshop was sponsored by the World Bank, the UK Department for International Development (DFID), PROFOR and the German Agency for Technical Co-operation (GTZ).
- Environmental Fiscal Reform for Sustainable Development and Poverty Reduction (held on 24-25 November 2003 in Berlin, and hosted by BMZ/GTZ).

The efforts and contributions of the sponsors and hosts of these workshops are gratefully acknowledged.

Table of Contents

Part I
The Political Economy of Environmental Fiscal Reforms (EFR):
Overview of Key Issues

Part II
Environmental Fiscal Reforms:
Political Challenges and Opportunities in Selected Sectors

ISBN 92-64-00868-3
Environmental Fiscal Reform for Poverty Reduction
© OECD 2005

Executive Summary

What is environmental fiscal reform?

"Environmental fiscal reform" (EFR) refers to a range of taxation and pricing measures which can raise fiscal revenues while furthering environmental goals. The international community has committed itself to the Millennium Development Goals (MDGs), including the overarching target of halving extreme poverty by the year 2015. To help achieve the MDGs, developing country governments need to mobilise revenue to invest in schools, health care, infrastructure and the environment.

EFR can play an important role in pursuing the MDGs of *"halving absolute poverty"* and of *"reversing the loss of environmental resources by the year 2015"*. Indeed, the UN Summits on Financing for Development and on Sustainable Development in 2002 recognised the potential contribution of EFR-related approaches. The latter stressed that poverty reduction and improved environmental management go hand-in-hand.

How can EFR contribute to poverty reduction and development?

EFR can contribute to poverty reduction *directly* by helping address environmental problems – such as water contamination and air pollution – that impact the poor. EFR can also help *indirectly*, by generating or freeing up resources for anti-poverty programmes in such areas as water supply and sanitation, or for pro-poor investments such as health and education.

EFR is an important part of the development policy tool kit. EFR approaches and instruments complement and strengthen regulatory and other approaches to fiscal and environmental management.

The multiple benefits of EFR

Fiscal benefits	Environmental benefits	Poverty reduction benefits
a) Revenue mobilisation	a) Pollution prevention and improved natural resource management	a) Improved access to water, sanitation and electricity services
b) Reduced distortions	b) Mobilisation of funds for investment in pollution control and safe disposal of waste	b) Mobilisation of funds for pro-poor investments (e.g. education)
c) Reduced drains on public finances	c) Mobilisation of funds for enforcement activities	c) Freeing up of financing to address environmental and other problems that affect the poor

What are the objectives of this document?

This document outlines the key issues to be faced when designing EFR. The objective is to provide insights and good practice on how development co-operation agencies can help developing countries take advantage of EFR approaches in their development and poverty reduction strategies.

Experience shows that despite its potential fiscal, poverty reduction and environmental benefits, EFR measures are constrained by political and institutional factors. Overcoming these factors requires thorough analysis of the political context, followed by effective management of the reforms as an inclusive political process. Accordingly, following a review of the instruments of EFR and related technical issues, the document focuses on the *political economy* and *governance aspects* of EFR. This includes an examination of the precondition for successful design and implementation, the various steps involved through the EFR policy cycle, the challenges to be faced at each stage and the main stakeholders involved. The role of donors in supporting EFR processes is also outlined. In Part II of the document, these issues are reviewed in relation to sectors of particular relevance for developing countries: forestry, fisheries, fossil fuel, electricity, drinking water and industrial pollution control.

Who is this publication intended for?

This document is primarily intended for decision makers and development policy experts in donor organisations who support developing country partners on fiscal or environmental policy issues, notably in the context of Poverty Reduction Strategies and Sector-Wide Approaches. They include *finance and environmental experts* as well as *sector specialists* working in such areas as forestry, fisheries, energy, water and transport. We also know from experience that the most effective assistance is that which supports country-led development programmes, and that builds on, rather than substitutes for, partners' own efforts. We therefore hope that the analysis and recommendations in this document will be of use to *policy makers* as well as to representatives of *civil society* groups and the *private sector in partner countries.*

Which EFR approaches apply to developing countries?

The range of EFR instruments is wide. Different EFR approaches and instruments are applicable to different sectors and issues. This publication covers instruments most relevant to developing countries. These include *a)* taxes on natural resource extraction such as fisheries and forests, *b)* cost recovery and pricing measures to improve access to basic services such as water and energy, *c)* taxes and subsidy reforms to discourage the use of environmentally damaging products, and *d)* taxes and fees to control harmful industrial pollution and waste.

EFR measures are feasible in most developing countries and indeed are applied by many. However, different measures may be more appropriate for different countries and sectors. While there can be no simple generalisation, they fall into the following broad groups:

- Natural resource pricing measures, such as taxes for forests and fisheries exploitation, are relevant for most resource-rich countries – which tend to be low-income countries.

- Reforms of product subsidies and taxes are applicable to most countries, but particularly energy producing countries where fuel subsidies are often high.

- Cost recovery measures, such as user charges on energy and water, are applicable to most countries but must be implemented carefully to protect the poor.

- Pollution charges are particularly relevant for rapidly industrialising middle-income countries where industrial pollution is a serious problem but administrative capacity is relatively strong.

Comprehensive approaches to development, such as sustainable development strategies, poverty reduction strategies and sector-wide approaches provide important opportunities for EFR. These approaches provide new opportunities to integrate EFR in the context of country-led national and sectoral development plans. Medium-term expenditure frameworks (MTEF), in particular, address directly issues relating to fiscal revenue collection and tax and pricing reforms. A number of Poverty Reduction Strategy Papers make reference to EFR approaches to address either fiscal or environmental objectives relevant to the poor.

What key principles should EFR design follow?

Taxing "bads" rather than "goods". It is generally preferable to tax actual pollutants (*e.g.* sulphur emissions) rather than inputs or products which are associated with the generation of pollution (*e.g.* fuel which contains sulphur). Similarly waste collection treatment charges should take account of actual volumes discharged and qualitative factors (such as concentration of toxic substances), and discriminate against the most harmful emissions. In some cases, however, discriminating approaches are not practical and more blunt instruments (such as taxes on inputs or outputs) must be used.

Setting the level of taxes. Ideally, from an environmental standpoint, a tax should be equal to marginal cost of abatement to provide incentives for actual pollution reduction. If it is lower than that, firms will generally prefer to pay the tax. The charges would thus operate primarily as a revenue raising mechanism with few pollution-reduction benefits. In practice, however, low levels of taxes or charges can help establish the principle that industries should pay for pollution and invest in research with new instruments.

Allocating the proceeds. Principles of public finance argue against allocated revenues from taxes to pre-determined purposes (earmarking). At the same time, allocating at least a share of the proceeds from environmental taxes to monitoring and enforcement activities (partial earmarking) can be justified. This may be a price worth paying for having predictable financing for environmental monitoring and enforcement activities, notably in countries where environmental agencies are poorly resourced or heavily dependent on donor assistance. It may also reinforce incentives to enforce environmental regulations and collect associated taxes and fees as well as, importantly, to help generate public support for EFR. Allocating some part of the proceeds from pollution taxes to the industry ("revenue recycling") may also be considered, in order to make the introduction of pollution charges politically more acceptable. In such cases, it is essential for the financial reflows to industry to discriminate between good and bad performers, and discourage continuous bad practice.

Respecting the principles of sound public finance. Revenues from EFR allocated to environmental purposes must be managed in line with internationally-recognised principles of sound public expenditures management, relating to fiscal discipline, efficient allocation of public funds, operational efficiency, accountability, transparency and comprehensiveness of the budget. Rigorous and regular evaluation is needed in this context. Where revenues are used to encourage investment in environmentally preferable equipment and methods, this should be strictly time-bound.

Matching instruments and implementation capacity. New instruments should be developed within the context of existing regulatory and institutional frameworks. Their type and scope must match the institutional capacity to implement them. Good mechanisms include pollution charges which can, for example, be collected and managed by environmental protection agencies, which are responsible for monitoring and enforcing existing laws. Pollution taxes "built in" to the national fiscal system and collected by fiscal authorities may be more complex to implement. China, for example, first experimented with the first approach for several years before moving to the second.

Building the capacity and credibility of environmental agencies. Monitoring agencies must collect accurate and timely information on industrial pollution flows, their origins and their impacts on, *e.g.* water and air quality. Partial earmarking of pollution taxes can help develop and maintain the capacity of environmental monitoring agencies and protect them from politically motivated attacks on their budget.

EFR implementation: the "political economy" dimension

The fiscal, environmental and poverty-reduction benefits of EFR can go hand-in-hand. But this is not automatic. There can also be trade-offs between various objectives which must be addressed explicitly. EFR requires careful policy design, taking account of issues relating to i) equity; ii) fiscal and environmental effectiveness; iii) administrative feasibility and efficiency and, iv) political feasibility. The "political economy" dimension is critical. Key aspects of policy design and implementation include:

- **Identifying likely winners and losers and understanding the perspectives and interests of affected stakeholders** – political leaders; poor and vulnerable groups; the private sector; the government; civil society groups; and the media. How the revenue raised (or freed up) as a result of reform is allocated is also crucial.

- **"Building in" compensatory measures.** Revenues from EFR can be allocated to poverty reduction and environmental management purposes. In some cases, time-bound support for affected communities and industries is necessary to secure political and public support for reform.

- **Building on public pressure.** Public awareness and participation invariably play a key role in building the necessary political will to enact and enforce pollution regulation, and overcome resistance of industry. A key requirement is the public disclosure of information on the health hazards from pollution as well as on industry's impacts on the environment. Such information must be made available to the public in ways that are easy to understand. In some countries "shaming the polluters" by publishing information on the performance of individual firms is an effective way to pressure them.

- **Helping firms comply with standards.** There are many low-cost ways for industrial producers to reduce pollution and waste. Compared with standard methods, "cleaner production techniques" use energy, raw materials and other inputs more efficiently, leading to cost savings. By focusing on prevention, cleaner production approaches reduce the need for "end of pipe" pollution control equipment.

- **Sequencing reforms and combining instruments:** Scheduling and announcing future increases of charges or taxes in advance and allocating the proceeds to help firms invest in pollution control can facilitate adjustment and reduce resistance to the introduction of charges. Gradual approaches also provide room for the necessary consultation with affected industries.

Table 1. **Poverty impacts of selected EFR Instruments**

Type of instrument	Potential impacts on the poor	Ways to enhance the benefits to the poor
Rent taxes (minerals, forestry, fisheries)	Generally positive if taxes are on commercial operators and some revenues used to benefit the poor.	Ensure that the poor are not negatively affected by commercial-scale harvesting; and that revenues intended for poor are not lost through corruption.
Petroleum excise taxes	Increased prices, especially of diesel, can increase cost of public transport and general cost of living – especially for remote communities.	Channel resources for improved public transport.
Electricity user fees	Raises prices for the poor, depending on the extent to which they are connected to grid (generally higher in urban areas).	Targeted subsidies for the poor ("lifeline tariffs").
Domestic water user fees	Raises prices for the poor, depending on the extent to which they are connected.	Targeted subsidies for the poor ("lifeline tariffs").
Irrigation user fees	May raise prices for the poor, depending on their access to irrigation services.	Targeted subsidies.
Increased prices for fertiliser and pesticides	Depends on access of poor for fertilisers and pesticides.	Targeted subsidies.

How can donors contribute?

The EFR policy cycle involves a number of linked and often over-lapping phases. Donors can play important roles at each stage:

a) Agenda setting stage

EFR must start with a clear identification and sound understanding of the issue to be tackled, including notably its impact and causes and its relative importance in view of the many pressing issues facing a country. **Donors** can play an important role in this regard by supporting conferences and round tables placing the problem on the agenda as well as the work of universities, research groups and international organisations in relevant areas.

b) Policy development stage

"Policy development" involves an assessment of the mix of instruments (fiscal and non-fiscal) which could be used to address the problem identified most efficiently, given existing socio-political and institutional conditions. It also includes identifying potential "winners" and "losers" from reform and possible compensatory measures. Through their support to various economic sectors, sustainable development strategies, poverty reduction strategies and sector-wide approaches, **donors** can encourage integrated, cross-sectoral policy reforms including in relation to EFR. This includes, in particular, ensuring that available opportunities for "win-win" EFR approaches are not missed.

Donors can also support those sectors of government (such as finance or environment ministries) who favour reform to overcome bureaucratic inertia or resistance from hostile ministries or agencies. They can also encourage transparency, access to information concerning

ENVIRONMENTAL FISCAL REFORM FOR POVERTY REDUCTION – ISBN 92-64-00868-3 – © OECD 2005

public finances, public participation, and accountability, which are key preconditions for sound policy development and, more generally "good governance". They can support capacity development, notably in the relation to the measurement of implicit taxes and subsidies and the quantification of impacts.

c) Dialogue, information dissemination and advocacy stage

Identifying a problem and possible solutions is not enough. Securing political acceptance and public support for EFR proposals often requires active advocacy, including through public awareness campaigns. Where the problems associated with corruption and patronage are serious, resistance to EFR and particularly around natural resources will be particularly strong and building strong alliances is vital.

Donors can contribute to such dialogues and awareness-raising, including through support to civil society groups. They can also support international sharing of experience and dialogue on EFR among developing country governments, international organisations and NGOs. Examples of this include current initiatives, for example, on the transparency of extractive industries (Extractive Industries Transparency Initiative – EITI) and illegal logging (Forest Law Enforcement Governance and Trade Process – FLEGT). Donor agencies can also make available information on reform experiences from OECD countries, bearing in mind that many of these also face various challenges in implementing successful EFR.*

d) Implementation stage

Implementation generally starts with public announcement of upcoming reforms preferably early on. It is important to give affected parties the time to prepare and adapt to the proposed changes. Gradual phasing-in of reforms is another way to reduce the transition costs, although it may increase the risks of loss of momentum.

Donors can play an important role by helping to finance the transitional costs of reform, in order to protect the poor from negative impacts or to overcome politically powerful blockages. This also includes supporting technical co-operation to help industries adjust to change (*e.g.* by switching to cleaner production techniques).

e) Building the credibility of monitoring and enforcement agencies

Credibility is essential to sustain support for reform, and confront challenges. This includes, for example, resistance from affected industries which have direct interests in portraying the environmental monitoring and enforcement agency as unprofessional, corrupt, or abusive. Hence, support to build the capacity of agencies to monitor and enforce compliance with applicable regulations is crucial. Environmental agencies must also be credible vis-à-vis ministries of finance. This includes, in particular, conforming to rules and principles of public expenditure management. This is particularly crucial when they are entrusted with the collection and management of taxes or fees or when the proceeds from environmental taxes and fees are earmarked and transferred to them for environmental management purposes. The capacity of fiscal authorities

---

* For instance, while various OECD countries have substantially extended the use of environmental taxes since the early 1990s, there are currently still many exemptions and tax rate reductions currently granted to sectors most exposed to international competition. For a more detailed discussion of these issues, see OECD (2004), *Implementing Sustainable Development: Key Results 2001-2004.*

to collect and administer existing taxes (including compliance monitoring and enforcement) may also need to be enhanced for proposals for fiscal reforms to be credible.

Donors can play important roles in providing technical assistance to develop partners' capacity in all these areas. In the case of natural resources, this may include the formulation of new management regimes as well as the formalisation and registration of property rights.

General principles to guide donors' actions in all these areas include:

1. Country ownership and sensitivity to the local context

In their policy dialogue on EFR, donors must take full take account of local conditions. Donors should encourage country ownership and not force the pace. Donor agencies also have to be sensitive to political challenges of EFR and attuned to the cultural and historic particularities. They should avoid imposing "blueprints", but instead provide financial, technical, institutional and political support to countries' own efforts. Strong country ownership, in turn, encourages donor harmonisation and will shield countries from excessive donor intrusion.

2. Readiness to act opportunistically and to be pragmatic

In a politically and economically volatile environment, it is crucial to take advantage of windows of opportunities as and when they arise. A new political leadership – especially if it has popular support – can be the catalyst for major policy shifts. Donors should be prepared to seize on the new opportunities for reforms which this creates.

It may also be necessary – and justified in some cases – to deviate from standard fiscal practice and to assign some part of tax revenues from EFR to a particular use in order to secure political will and/or public support for important reforms. In such cases, donors can play an important role in working with fiscal authorities to ensure that appropriate safeguards are put in place to ensure consistency with efforts towards fiscal consolidation.

3. Harmonisation, alignment and policy coherence

In EFR, as in other areas, donor support should be harmonised – with donors working together to avoid duplication of missions, studies or institutions. And it should be aligned behind country-owned and led strategies and procedures. Accordingly, EFR measures should be conceived within the context of PRSPs, sustainable development strategies, medium-term expenditures programmes and sector-wide approaches. More generally, donors should also ensure that their efforts toward poverty reduction, environmental sustainability and sound economic and fiscal management are mutually supportive.

Coherence between development policies and other policy areas is also vital: donor countries can undermine the objectives of their development co-operation policies by inconsistent policies in other areas. Donor agencies should strive to raise this issue within their governments to draw attention to this risk and try to minimise it. For example, donor agencies should raise awareness in their governments of, *e.g.* agricultural policies which distort world market prices, or fisheries agreements with developing countries which run against environment and poverty reduction objectives.

Busy Readers' Guide to this Document

PART I

The Political Economy of Environmental Fiscal Reform (EFR): Overview of Key Issues

Part I of this DAC Reference Document outlines key issues to be faced when designing Environmental Fiscal Reform (EFR). EFR refers to a range of taxation and pricing measures which can raise fiscal revenues while furthering environmental goals. Specific sectors reviewed include forestry, fisheries, fossil fuel, electricity, drinking water, and industrial pollution. The objective is to provide insights and "good practice" on how development co-operation agencies can help developing countries take advantage of EFR approaches in their development and poverty reduction strategies.

Following a review of the instruments of EFR and related technical issues, the document focuses on the political economy and governance aspects of EFR. This includes an examination of the preconditions for successful design and implementation, the various steps involved through the EFR Policy Cycle, the challenges to be faced at each stage and the main stakeholders involved. The role of donors in supporting EFR processes is also outlined.

The Political Economy of Environmental Fiscal Reform (EFR): Overview of Key Issues

ISBN 92-64-00868-3
Environmental Fiscal Reform for Poverty Reduction
© OECD 2005

PART I

Chapter 1

Introduction: Environmental Fiscal Reforms and Poverty Reduction

What is EFR?

"Environmental fiscal reform" (EFR) refers to a range of taxation and pricing measures which can raise fiscal revenues while furthering environmental goals. This includes taxes on natural resource exploitation or on pollution. EFR can directly address environmental problems that threaten the livelihoods and health of the poor. EFR can also free up economic resources or generate revenues that can help to finance access of the poor to water, sanitation and electricity services.

Different EFR approaches and instruments are applicable to different sectors and issues. These instruments do not substitute for but complement and strengthen regulatory and other approaches to fiscal and environmental management. EFR instruments should therefore be thought of components of fiscal and environmental policy mixes, not as "stand alone" policy packages. Similarly, it is only one of the ways through which fiscal authorities can raise additional revenue. This document covers instruments relating to natural resource extraction, environmentally damaging products, air and water pollution and cost recovery instruments for environmental management infrastructure.[1]

How can EFR contribute to poverty reduction?

EFR can contribute to poverty reduction directly by helping address environmental problems that threaten the health and livelihoods of the poor such as water contamination and air pollution. EFR can also help generate resources, notably to help achieve Millennium Development Goals (MDGs) and for other pro-poor programmes. For example, it can help finance investment in infrastructure critical for the poor, such as water supply and sanitation. It can also help finance other pro-poor investments or services in areas such as education and health.[2]

How does EFR fit in the current international development context?

There is renewed interest in EFR for several reasons. The international development agenda is now strongly focused on *the Millennium Development Goals* (MDGs). These include the goals to eradicate extreme poverty and hunger and ensure environmental sustainability.

The *World Summit on Sustainable Development* (WSSD) held in South Africa in 2002 re-affirmed the MDGs and stressed that poverty reduction and improved environmental management go hand in hand. The WSSD Plan of action calls upon governments to "... *continue to promote the internalisation of environmental costs and the use of economic instruments, taking into account the approach that the polluter should, in principle, bear the costs of pollution, with due regard to the public interest and without distorting international trade and investment*".[3]

The *Financing for Development* Conference, held in 2002 in Monterrey, also emphasised the importance of mobilising domestic resources for development including through efficient tax systems. EFR is one of the ways to increase domestic resource mobilisation.

Current approaches to development also emphasise the role of the state in establishing its strategy for poverty reduction and in financing poverty reducing investments. The *Poverty Reduction Strategy* (PRS) process, launched in the late 1990s, reflects this approach. It is strongly encouraged by development co-operation agencies who are keen to shift from project-by-project approaches toward more comprehensive strategic frameworks to guide development co-operation efforts to reduce poverty.

Poverty reduction strategies provide important opportunities to integrate EFR approaches into national and sectoral development plans. Medium-term expenditure frameworks (MTEF), in particular, address directly issues relating to fiscal revenue collection and tax and pricing reforms. A number of Poverty Reduction Strategy Papers (PRSPs) make reference to EFR approaches either to address fiscal or environmental objectives relevant to the poor.

Governance, transparency and public participation

The design and implementation of development policies entails a number of trade-offs. Political, institutional and social issues play a central role. EFR is no exception. Issues of "good governance", transparency and public participation in decision-making play a central role. This is particularly important in relation to the exploitation of natural resources, where the scope for corruption, rent-seeking and political patronage is significant.

The feasibility of EFR measures thus depends on country-specific factors such as respect for the rule of law, ability to enforce property rights, possibilities of legal action by citizens or NGOs in case of violation of existing laws, freedom of information, or capacity and authority of fiscal and regulatory institutions. These governance and institutional factors, which also condition traditional regulatory approaches, will generally vary across sectors in a single country. The merit of EFR approaches will vary accordingly. For example, a very poor country endowed with a rich natural resource base may give priority to developing the institutional means to implement EFR approaches in natural resource extraction while sticking to traditional regulatory approaches with respect to other environmental management issues.

In many countries, moves towards democratisation, liberalisation, decentralisation and increased public participation in decision making are increasing the political feasibly of EFR. The merit of EFR approaches relative to other ways of raising revenues will also depend on cost effectiveness considerations.

EFR in developing countries: lessons from past experience and future prospects

In many developing countries, economic and fiscal priorities have, in the past, been the key driving force behind environmental fiscal reforms. For example, the debt crisis and deteriorating balance of payments positions affecting many developing countries since the early 1980's have triggered profound structural reforms, often with financial support from external development agencies. It became urgent to restore macroeconomic stability, improve revenue mobilisation and increase efficiency of public spending. "Structural adjustment" has often included pricing and subsidy reforms, increased cost recovery of publicly provided services, privatisation of many industries, and increased trade liberalisation.

Although environmental management was not the main motive, these reforms have often had beneficial environmental impacts. Examples of this include the reduction of subsidies on pesticides, water, and energy, which have encouraged more rational use and reduced the negative environmental impacts. Both China and Russia significantly increased energy prices during the 1990s with major environmental benefits. However there have been social costs when, for example, increased prices of basic goods and services consumed by the poor were not accompanied by compensatory measures. This risk is now well recognised.

While EFR measures are feasible in most developing countries, different measures may be more appropriate for some countries and sectors. Even though there can be no simple generalisation, different instruments will fit different situations:

- Natural resource pricing measures, such as taxes for forests and fisheries exploitation are relevant for most resource-rich countries – which tend to be the low income countries.

- Reforms of product subsidies and taxes are applicable to most countries, but particularly energy producers where fuel subsidies are often high.

- Cost recovery measures such as user charges – on energy and water – are applicable to most countries – but must be implemented carefully to protect the poor.

- Pollution charges are particularly relevant for rapidly-industrialising middle-income countries where industrial pollution is a serious problem but administrative capacity relatively strong.

Experience in OECD countries and economies in transition

In many OECD countries, increased use of fiscal instruments for environmental management has been rooted in the recognition that environmental challenges are beyond the remit of environment ministries alone and that fiscal instruments can reinforce traditional regulatory approaches in helping curb environmentally damaging consumption and production patterns.

The critical role of prices and markets in promoting resource efficiency was sharply illustrated by the experience in former socialist economies. There policies based primarily

---

Box 1.1. EFR in OECD countries

OECD countries have made considerable progress on EFR, notably in relation to transport-related energy taxes. In these countries, the revenue from environmentally related taxes averages roughly 2% of GDP. However, there is considerable room for further progress: For instance, subsidies to the energy sector, and notably on coal, are still considerable despite reductions over the past two decades. Agricultural subsidies remain at high levels, often with negative environmental effects, (*e.g.* output-increasing policies have encouraged the expansion of farming on to environmentally fragile land). These issues are not covered here as the focus is on developing country experience – and there are many other OECD publications on industrialised country experience.

Source: Sustainable Development: Critical Issues (OECD, 2001).

---

on regulations ("command and control") and heavy subsidisation of goods such as energy and water had induced major environmental problems without commensurate economic benefits.

While experience with market-based instruments (MBIs) for environmental management has not been uniform across OECD countries, there have been significant successes (OECD, 2001). In addition, more recently there has been more emphasis on the revenue benefits of environmental taxation – often known as "ecological tax reform" in order to shift from taxing "goods" like labour to taxing "bads" like pollution and waste. Countries such as Germany, Sweden and the United Kingdom have experimented such approaches.

EFR: Potential benefits and tradeoffs

Potentially EFR can yield multiple dividends in terms of financial resource mobilisation environmental management and poverty reduction.

a) Fiscal benefits

Developing countries face formidable challenges in raising revenue to finance essential public services. As a large share of economic activities are small-scale or informal, the collection of income, sales and corporate taxes is very difficult. Tax evasion and arrears (notably by state-owned enterprises) and corruption are widespread. Many countries thus rely heavily on import or export taxes, which are the easiest to collect, but increasingly incompatible with international trends towards trade liberalisation (IMF, 2001). Many developing countries are thus seeking to strengthen their tax systems, focusing on simple, transparent and broad-based taxes. EFR is one of the ways to do this, though by no means the only one.

In areas such as natural resources extraction and energy consumption, EFR has the potential to generate significant additional revenue. For example, in 1999 Mauritania doubled the revenue it earned from its fisheries; it represented about 30% of total tax revenues. In many countries, subsidies on energy represent a significant drain on public finances. For example, the power subsidy to Indian agriculture – which benefits large farmers – costs the country USD 6 billion a year – and is twice the national health budget.

EFR can also contribute to minimising distortions in tax systems and supporting economic growth in general. It can be viewed within the overall context of reform in developing countries' move towards incentive-based market economy. In all cases, the merit of EFR relative to other revenue raising approaches must be assessed by reference to cost-effectiveness, administrative feasibility and other pertinent factors.

b) Environmental benefits

It is difficult and costly to enforce environmental regulations when prices are sending out the wrong signals. Measures such as taxes on pollution or environmentally harmful products, such as pesticides, or on the extraction of natural resources can create *direct incentives for minimising pollution and resource waste* by indicating through the price mechanism the true cost of pollution, and the scarcity value of natural resources. For example, rises in the price of energy in China stimulated major improvements in energy efficiency. In Indonesia, removing pesticide subsidies saved the government over USD 100 million a year while improving the environment and the safety of agricultural workers.

In addition to providing financial incentives (or reducing the disincentive) to comply with environmental regulations, EFR instruments can *generate the funds to cover the costs of monitoring and enforcement* activities, which are in many countries under-funded.

Fiscal instruments can also make *industrial polluters pay for the cost of cleaning up*. This includes, for example, levies to finance waste disposal and treatment infrastructure and services. This frees-up revenues for governments who often provide such services for free. In Mexico, for example, waste management is largely financed by municipalities' general budgets with around 20% only of the operating costs recovered from user charges.

c) Poverty reduction benefits

As highlighted in the WSSD Plan of action, among others, the poor suffer disproportionately from the impacts of many forms of environmental degradation. EFR can contribute to poverty reduction *directly* by addressing environmental problems that matter to the poor. For example, reduced air pollution as a result of EFR measures has direct health benefits for the poor who are disproportionately exposed to harmful pollutants. Similarly, EFR measures to reduce forest degradation by commercial logging firms can directly benefit poor communities who rely on forests for their livelihood. EFR can also generate resources to expand infrastructure in areas such as, water supply, sanitation and energy, to poor rural communities who do not have access to such basic services. EFR can also contribute to poverty reduction *indirectly*, by freeing up budget resources which can be allocated to pro-poor investments in areas such as education and health.

Table 1.1. **The multiple benefits of EFR**

Fiscal benefits	Environmental benefits	Poverty reduction benefits
a) Revenue mobilisation	a) Pollution prevention and improved natural resource management	a) Improved access to water, sanitation and electricity services
b) Reduced distortions	b) Mobilisation of funds for investment in pollution control and safe disposal of waste	b) Mobilisation of funds for pro-poor investments (e.g. education)
c) Reduced drains on public finances	c) Mobilisation of funds for enforcement activities	c) Freeing up of financing to address environmental and other problems that affect the poor

Formulating EFR: key issues to consider

The fiscal, environmental and poverty-reduction benefits of EFR outlined above can go hand in hand. But this is not automatic and requires careful policy design. There can also be trade-offs between various objectives which must be addressed explicitly. The design of EFR should therefore consider the following aspects:

a) Equity

The equity and poverty impacts of EFR measures depend on who is directly and indirectly affected by the measures. For example the removal of an untargeted subsidy on irrigation water will tend to affect the heaviest users, who are generally not amongst the poorest. Farm labourers, however, may also be indirectly impacted. For some heavily polluting inputs, such as diesel fuel, however, the poor are heavy users and will be heavily impacted by EFR measures.

How the revenue freed up by the measures is used is also critical. It is often possible to allocate at least part of this revenue for targeted measures to compensate the poorest or those most affected by the measure. A detailed analysis of these effects and how different social groups are impacted is needed to assess the likely overall impact of EFR measures and to design compensation measures.

b) Fiscal versus environmental effectiveness

The balance between the revenue-raising and environmental benefits of EFR measures will depend on how the reforms are designed. For example, a pollution tax may be set too low to induce change in production techniques but it may be successful in raising revenue. At the opposite end, a tax on a relatively unessential or easy to substitute but highly polluting input may lead to its complete phase out, yielding significant environmental results but minimal revenue. In most OECD countries, however, road fuel taxes contribute significantly to both fiscal and environmental objectives.

c) Administrative feasibility and efficiency

Implementation capacity is one of the factors conditioning the successful implementation of EFR. The ideal, from an environmental point of view, is to target the environmental "bad" as closely as possible (e.g. a particular harmful pollutant). Given weak environmental monitoring capacity, however, it may be simpler and ultimately more effective to use a relatively blunt instrument such as a tax on products which generate those harmful pollutants in the course of their production. For example, a tax on fuel may be the best way to reduce the emission of a range of harmful pollutants which are difficult and costly to monitor. This works well for "dirty fuels" where harmful emissions and fuel consumption are closely linked. Pragmatic factors will determine how such trade-offs are addressed. The legal basis for environmental monitoring and enforcement efforts as well as the credibility of the enforcement agency are other key issues determining the choice of EFR instruments.

d) Political feasibility

Most EFR measures will entail losses for some. Energy-intensive industries, for example, stand to lose from energy taxation measures. Issues related to corruption and political patronage are particularly critical in the area of natural resource extraction (e.g. timber or minerals) insofar as those resources are used as sources of wealth for political elites. Clearly, pro-poor and environmentally beneficial reform proposals can face severe political opposition.

When those affected adversely by reforms are very poor or, alternatively, politically influential, it may become necessary to allocate part of the revenue back to them in order to ensure that the measure is adopted. There can therefore be trade-offs between equity, environmental and fiscal effectiveness and political feasibility considerations. All of these should therefore be considered an integral part of policy design.

The next chapter will provide an overview of the main instruments of EFR and related "technical issues". This will pave the way for taking up the most critical issue: the "political economy" dimension of EFR and the precondition for successful implementation.

Notes

1. Tradeable pollution permits and deposit refund schemes are not covered here because, in general, they are not intended to raise revenue.

2. The links between poverty, gender and environment, and their policy implications are examined in greater detail in OECD/DAC (2001) ("Poverty-Environment-Gender Linkages", DAC Journal 2001, No. 4).

3. WSSD Plan of Action, paragraph 18b.

ISBN 92-64-00868-3
Environmental Fiscal Reform for Poverty Reduction
© OECD 2005

PART I

Chapter 2

The Instruments of EFR

Environmental fiscal reform covers a broad range of instruments that deal with sustainable use of natural resources and the control of harmful pollution. In general, an EFR measure is not implemented in isolation but as part of a package including notably complementary regulatory measures. Here we review the various instruments of EFR and some of their main uses and features. The focus is on instruments which can be used to both raise revenue and improve environmental management. Instruments with limited revenue-raising potential are not considered. Depending on conditions in a given country and sector, different measures will be relevant. In Part II below, these instruments will be reviewed in the context of specific sectors.

Taxes on natural resource extraction

On average, developing countries tend to be far more natural resource-dependent than OECD countries. Natural resources provide inputs to many industries and play critical roles as sources of livelihoods, especially for the poor. Many of these resources are legally owned by the state and provide important sources of revenue when they are commercially extracted.[1] The state takes a share of the extractors' profits – or, in economic jargon, "economic rent". This applies both to renewable (*e.g.* forests and fisheries) and non-renewable resource (*e.g.* minerals). Sound management of natural resources is therefore an important factor of sustainable economic growth and poverty reduction and is essential for continued revenue generation.

Taxes on large-scale commercial forestry and fishery extraction already generate considerable revenue in many low-income countries. In Guinea Bissau, for example, revenues from leasing out fishing resources accounted for around 30% of government revenue between 1993 and 1999. Some forest-rich countries – such as Cambodia and Cameroon – have undertaken reforms to increase the fiscal revenues from the forestry sector. This is recognised in some PRSPs (*e.g.* Ghana, Mauritania, Cambodia). However, there often is scope for further increasing these revenues and improving resource management. The design of EFR in the forestry or fisheries sectors is complex, including institutional capacity and risks of corruption, and there are potential trade-offs in the different instruments. These issues are reviewed below (see Chapters 5 and 6)

Product subsidies and taxes

a) Product taxes

When the production, use, or disposal of certain products creates health-threatening pollution or hazardous wastes, their use can be discouraged through taxes. Products submitted to such taxes include energy products, fertilisers, batteries, pesticides and tobacco.

Taxes on environmentally harmful products can both reduce their use, thereby improving the environment, and generate fiscal revenue. The balance between these two effects depends on how responsive (in economic jargon: price-elastic) demand is to a price increase. If demand is not very price-sensitive and the product continues to be purchased, then the tax can yield considerable revenue but more modest environmental benefits. If, on the other hand, demand is very responsive to a price change, then use of the product will decline sharply in response to the tax, but little revenue will be generated. This will happen when substitutes are readily available.

A product tax must therefore take account of the scope for switching to substitute products. If those substitutes are equally harmful, the tax will yield neither fiscal nor environmental benefits. In many cases, different tax rates are applied precisely to encourage switching towards less environmentally harmful substitutes. Thus Thailand, like many OECD countries, has used differential tax rates to promote unleaded petrol to encourage switching (see Box 7.3 below). For similar reasons, China taxes high-sulphur coal more heavily than low-sulphur coal.

In practice, product taxes can both yield significant revenues and improve the environment. In OECD countries taxes on the purchase or use of motor vehicles and fuels, including those on petrol and diesel, currently generate most (around 90%) of the revenue from environmentally related taxes.[2] At the same time, they play an important role in containing congestion and controlling air pollution from transport, while stimulating fuel efficiency. Such taxes are also significant in many developing countries. In Sub-Saharan Africa, excise taxes on petrol and diesel fuel generated an average of 35% of government revenue in 1990. However in many developing countries, taxes on environmentally harmful fuels, such as diesel, and on pesticides and fertilisers, are low.

The application of a tax on inputs such as energy can create competitiveness concerns on the part of industry if competitors in other countries do not face such taxes. In OECD countries many industries, as well as agricultural activities, are exempted from such taxes and the burden is concentrated on households. In the developing countries context, particular attention is needed to assess the impact of such taxes on the poorest (see Chapter 7 below).

b) Product subsidies

The fertiliser subsidy scheme has been streamlined so as to release funds for a targeted, voucher-based farm inputs support scheme for small producers. (Sri Lanka PRSP: 35)

Subsidies are harmful to the environment if they lead to higher levels of damage than would otherwise occur. In many developing countries, governments do not tax but actually subsidise certain environmentally harmful products, thereby encouraging over-use. Examples include subsidies for polluting fuels, pesticides, fertilisers and water.

Some subsidies are explicit and transparent: the government pays producers or consumers of the products in question and the payments appear in the budget as expenditures. In this case, it is possible to assess whether they actually benefit the poorest. Other subsidies are hidden, taking the form of tax concessions, regulatory exemptions, grants or loans at below market interest rates, price support measures, preferential depreciation allowances and other measures. As such they are not readily identifiable and it is difficult to estimate their magnitude and actual impact. Hidden subsidies often take the form of free or below-cost provision of goods or services, such as water or electricity. This is addressed below under "user charges".

The potential fiscal benefits of subsidy reform are clear as these subsidies often represent a large drain on public finances, depriving other sectors of the economy of budgetary resources. For example, Indonesia removed pesticide subsidies in 1986 and saved USD 100 million a year.

As in the case of taxes, the environmental benefits of subsidy reduction or removal depend on the cost of alternatives products – and their associated environmental impacts. In Indonesia's reform of their pesticide subsidies, for example, farmers shifted to integrated pest management – which was less risky, and more cost-effective. The environmental benefits of reducing subsidies on fertilisers are less clear-cut. They depend on current levels of use, the marginal returns to fertiliser use, and trade-offs between intensification (getting more production on each hectare) and extensification of agriculture (i.e. expanding to previously uncultivated land, sometimes by clearing forests). In Sub-Saharan Africa, fertiliser subsidies may have positive impacts given that soils there have suffered major losses in nitrogen and phosphorous nutrients in recent decades, and rates of fertiliser use are very low relative to the global average. In this context, fertiliser subsidies may be environmentally beneficial, helping to reverse net nutrient depletion, reduce soil erosion and increase yields (Aune and Oygard, 1999; Munasinghe and Cruz, 1997, both cited in OECD, 2002).

Taxes on polluting emissions

The unintended harmful side effects of economic activities, such as the release of health-threatening pollution, are not reflected in the market price of goods and services. Taxes on such "bads" are a direct way to help correct this. Air pollution and effluent discharge fees are generally set proportional to the emission levels of certain pollutants, thereby encouraging pollution reduction, while raising revenue. This approach, which requires an ability to monitor the quantity and quality of emissions, at least from large industrial sources, has been widely used in many developing countries, including China and Colombia. When pollution monitoring capacity is weak, blunter but simpler instruments such as product taxes, which work indirectly (see above) are preferable. Taxes on pollution are addressed in detail in Chapter 10 below.

User charges/fees

... another common measure in all four sub-sectors [energy, transport, water supply and communications] is tariff reform. The level and structure of tariffs is being revised to enable operating companies to become financially viable. Eventually, tariffs will be set to achieve full cost recovery. Affordability of services to the poor will be addressed either within the tariff structure or through separate targeted measures. (Tajikstan PRSP: 44)

User charges or fees are payments in return for the provision of a service. Examples include tariffs for electricity, water and sanitation, waste collection or disposal and industrial wastewater treatment. User charges do not generally go to the Treasury into the consolidated budget. Their primary aim is not to raise revenue as such but to cover the actual costs of providing a service. In the absence of user charges, costs have to be covered by the government's budget, which is already under stress. Frequent budget shortfalls lead to poor service.

The introduction of charges for services that were previously under-priced or provided free of charge can allow a needed service (e.g. water supply, electricity or industrial wastewater treatment) to be provided on a sustainable basis and/or release funds for other uses. It also has environmentally beneficial effects. Water and electricity tariffs encourage more efficient use, while fees for effluents or toxic waste treatment stimulate reduction in pollution and waste generation at the source.

At first sight, charging for essential goods such as water and electricity may be expected to hurt the poor the hardest. In practice, however, the poorest most often do not benefit from the free provision of those services, as they tend to not even be connected to the networks. The poorest must procure water and energy in other ways (water vendors and solid fuels) at higher costs. When the introduction of tariffs does hurt the poor, there are ways to provide targeted compensation or other types of protection. Issues relating to subsidy reform in the energy and water sectors are addressed in Chapters 7 to 9 below.

"Comprehensive" EFR

"Comprehensive" (or "cross-sectoral") EFR refers to approaches that build-in environmental considerations in macro- or sectoral-level tax policies, through reforms to instruments such as corporate taxes, depreciation allowance and others, in order to support broad fiscal reform objectives while providing an environmental orientation to the general tax structure. In OECD countries the focus of comprehensive EFR has been on introducing taxes on environmental "bads" to allow a reduction of other "distorting taxes", such as income taxes or taxes on labour, thereby improving the environment and making the tax system more efficient. This is called the "double dividend". In many cases, revenue neutrality of reform has proved to be a key factor of political feasibility. Leaving aside the merits of this approach, it can be considered of less relevance to most developing countries, where income taxes are still relatively low. Although there are exceptions to this, for example in South Africa, comprehensive EFR will not be specifically covered in this document.

Box 2.1. **"Ecological tax reform" in Germany**

In Germany, EFR was first conceptualised in 1983, and reached the political agenda during parliamentary election campaigns in 1990, when one Party prepared a concept based on raising energy taxation, and simultaneously reducing the tax burden for low income earners and raising transfers to pensioners. After a long delay, due notably to the reunification of Germany, EFR was again examined in 1998, after a government change. This generated fierce political debates. Potential losers from the reform, including energy-intensive industries (chemical industry, steel producers etc.) as well as employers associations and labour unions were resisting the introduction of EFR. In contrast, labour-intensive sectors and firms, especially service industries, were more open to reform. In 1999, the parliament agreed upon the "Law on the Introduction of Environmental Tax Reform". This reform i) raised taxes on gas and oil products; ii) raised taxes on electricity; and iii) reduced social security contributions. Contributions of both, employee and employer, to the public pension system were lowered significantly, thus reducing the cost of labour. The reform was designed as a stepwise introduction; until 2005, the initially low taxes will be raised annually, while the contributions to the public pension systems will be lowered.

Notes

1. Taxes on small-scale resource extraction, such as subsistence fishing or wood collection, generate much less revenue and can be regressive on the poor – so we do not cover them here.

2. OECD (2001a) defines "environmentally related tax" as any compulsory, unrequited payment to general government levied on tax-bases deemed to be of particular environmental relevance.

ISBN 92-64-00868-3
Environmental Fiscal Reform for Poverty Reduction
© OECD 2005

PART I

Chapter 3

Designing Reforms – The Politics of Making Trade-offs. From Policy Formulation to Actual Implementation

Proposals for EFR often focus on "what" needs to be done, rather than on "how" to actually make it happen. But actual implementation is often slow. Attributing this to "lack of political will", "lack of capacity" or "lack of resources" does not help much. Instead, the formulation of EFR proposals must from the outset take explicit account of the context in which they will have to be implemented, in order to identify likely constraints, and find ways to alleviate them. This includes, in particular, analysing the underlying political, social and administrative constraints. For example, proposals for the reduction or phase-out of certain subsidies must be framed in the context of political feasibility and allow for compensatory measures for affected sectors or groups. EFR proposals must also be consistent with and, if possible, reinforce other reform efforts, in areas such as finance, public administration and others.[1] This chapter will focus on the political economy dimension of EFR.

Balancing objectives: a political process

The objectives of EFR generally include raising public revenue, improving environmental management, and reducing poverty – with different groups attributing different degrees of importance to each objective. In trying to reach these goals, there might be win-win options, with more than one objective being achieved. For example, a product tax on energy may both raise revenues and help improve air quality, with significant health benefits to the poor. But there will also be implicit trade-offs between different EFR objectives. For example, a "pollution tax" set too low to encourage meaningful reduction in pollution may actually serve as a revenue-raising device. It may, however, be used as a signal to industries, and help to pave the way for subsequent increases.

In OECD and developing countries alike, the relative priority between different policy goals is determined by the political process. This includes inter-ministerial and cabinet discussions, parliamentary debates over the budget, back-room lobbying as well as public debates including through the media. It will imply confronting various perspectives. Below, we examine the key factors to be considered, when formulating EFR proposals and steps that can be taken to manage the reform process. Many of the issues raised are also relevant for reforms other than EFR.

Contextual factors for reform

1. Characteristics of the problems addressed by the reform proposal

EFR can address a variety of environmental problems, linked to different economic sectors. The structure of the proposal and its chances of success will depend on some key features of the problems addressed. These factors also determine the potential for mobilising political support for reforms. Relevant questions include:

Are the impacts of the environmental problem visible and immediate?

The more visible and immediate the problem, and the clearer the links to human health, the easier it is to mobilise support for addressing it. Many forms of air pollution are readily observable and the negative impact on health easily perceived or understood by lay persons. In contrast, problems such as ground water pollution or ozone layer depletion are invisible to the naked eye, or become apparent only with a time lag. In other cases as in the case of ozone layer depletion, they are related to human health in complex ways.

Can the problem be clearly linked to specific causes?

In many cities, air pollution can readily be linked to emissions from large-scale power plant and motor vehicles. In contrast, water pollution often results from a multitude of sources, both industrial and households. In general, problems linked to a narrow set of "causes" or "actors" are more readily tackled.

Are "technological fixes" or alternatives available?

Some environmental problems can be solved by technical measures. For example, fitting catalytic converters to automobiles can reduce harmful emissions, as can changes in the chemical formulation of fuels. If technological fixes or substitutes are readily available, it may be more difficult to generate support for fiscal options (i.e. taxing polluting fuels), and regulatory approaches (e.g. imposing the use of pollution control devices or regulating fuel quality) may be preferred, even where they are more costly and less effective. In general, though, fiscal and regulatory measures work best in combination.

2. Factors linked to the socio-political context

The historical, socio-cultural, economic and political-institutional context are key factors conditioning the scope for reform. For example in some countries, the perception of water as a non-economic good is deeply rooted, making it more difficult to introduce water pricing. Important elements of the "economic context" for EFR include historically-rooted patterns of land ownership, the degree of concentration in key industries as well as the degree of dependence on certain energy sources. Critical features of the "political context" include the nature of the governance system (patronage-based vs. democratic and based on the rule of law), the existence of transparent and participative decision making processes and the capacity of the bureaucracy. Other important factors include degree of literacy, existence of a free press, and availability of political space for open debate and confrontation among different stakeholders.

3. Factors linked to circumstances

Policy development and implementation processes are neither linear nor predictable: personalities, political or even accidental events can be important catalysts for major policy shifts. A newly elected popular leader can often launch major reform efforts. Major disasters like Chernobyl or Bhopal have triggered considerable policy advances in the field of environment, as have positive events such as UN Summits on environment, development and sustainable development (Stockholm 1972, Rio 1992 and Johannesburg 2002). Technological innovations can reduce the cost of addressing an existing problem and increase the willingness to tackle it although its impacts as such have not increased.

4. International agreements

In the context of growing international trade and awareness of the trans-boundary or global consequences of many environmental problems, international agreements, regulation and institutions influence national policies. For example, WTO rules limit countries' ability to impose certain forms of tariffs while the Montreal Protocol on Substances that Deplete the Ozone Layer provides for the phase-out of certain chemicals.

Identifying winners and losers: the transmission channels of reform

Most reform processes create winners and losers, at least in the short term. Identifying winners and losers is a critical part of EFR policy design, notably to build in well targeted compensatory measures. Public debates over impacts of reforms are influenced by interest groups' perceptions and biases as well as how they are presented by the media. Powerful interest groups (*e.g.* industry) often deliberately exaggerate the negative impacts to undermine reforms.

Reforms produce different effects in the short, medium and long terms. Perceived "winners and losers" will accordingly change over time. Increased electricity prices may have significant negative impacts for some households in the short term but could have longer-term benefits (through, *e.g.*, improved reliability of service). The benefits of adjustments to the change (*e.g.* switching to energy-saving devices and building materials) also accrue gradually. In some cases, the same households will experience these different impacts over time.

Likely winners and losers from reform can be identified by reference to the "transmission channels" of reform and their implications on different groups. These include:

- **Prices.** They determine real household purchasing power through direct effects on consumption (if households pay more for water) and indirectly through effects on production (if industries pay higher prices for certain inputs, they will pass on some of these increases onto consumers).

- **Employment** (informal or formal): It provides the main source of income. Some policies may, for example, shift demand for labour across industries or firms within an industry. For example, energy-intensive sectors may contract in response to increased energy prices, while producers of energy-efficient equipment or materials may expand.

- **Access to goods and services** (public or private). EFR can have direct impacts on households. For example, if water tariff increases allow the expansion of the network, it has direct benefits for those previously not connected.

- **Assets values** (financial, physical, natural, human or social). For example, reduction in traffic congestion and air pollution may increase the value of housing and land in impacted areas.

- **Transfers and taxes** can impact households. Increased cost recovery on publicly-provided services can provide room for reduction of taxes and/or free up government resources for other spending.

In practice, the identification of winners and losers from reform can be done through *expert analysis*, *e.g.* based on statistical data and/or surveys regarding, *e.g.* consumption patterns, exposure to pollutants, reliance on certain resources for income and the like. It can also be done through *participatory approaches*, with different stakeholders expressing their perceptions of how they believe they will be impacted (or a combination of both).

Perspectives and interests of key stakeholders

An understanding of how key stakeholders are likely to view a reform proposal and the incentives they face to support or undermine the reform process is critical to minimise opposition and maximise support. It is important to note, however, that most categories of stakeholders are far from homogeneous. The business sector, for example, is heterogeneous. Even within a single economic sector, there will generally be important differences across companies, with regard to, for example, energy efficiency, export competitiveness, and the use of advanced technology. Similarly, in governments, different ministries and even directorates within ministries pursue different objectives and are associated with different constituencies.[2]

It must also be noted that relatively small and unrepresentative but well-organised interest groups can exert disproportionate influence over policy, even in democratic systems. When they stand to lose, such groups can undermine reforms supported by a broad majority. For example, large-scale industries benefiting from free access to waterways to dispose of their wastes are often organised and well-represented politically and have ample means to impede reform efforts. Conversely, those affected by water pollution, while far more numerous, are disseminated throughout all sectors of society and are not easily mobilised around the issue. Bearing this in mind, this chapter will present a general overview of the perspectives of various groups.

a) Poor and vulnerable groups

Poverty reduction is a central objective of development co-operation. Accordingly, this publication focuses on the interest of the poor and vulnerable groups. But "the poor" are not a homogenous group. There are differences in gender, location (*e.g.* urban *vs.* rural), employment sources and many others. Analysis of EFR in different sectors, in Chapters 5 to 10 below, will identify the impacts on the poor in sector-specific circumstances.

b) The private sector

The private sector, which includes enterprises in different sectors and of very different size and ownership type ranging from the informal sector, small and medium enterprises (SME), large indigenous enterprises and foreign investors, will generally be affected by EFR reforms but in very different ways. Generally, industry will be affected by increases in the costs of key inputs, notably energy – and will often argue that this will lead to a loss of competitiveness. But investors in liberalised energy markets would have a different point of view, as would companies specialised in producing or selling energy-efficient technology, which would benefit through increased demand for their products from EFR.

c) The government bureaucracy[3]

In general, the **ministry of finance's** main concerns regarding EFR will be integration with the existing fiscal framework, the potential impacts on the budget (notably the potential for raising extra revenue) and the administrative feasibility of reforms. **Sectoral agencies** responsible for energy; water; agriculture, trade and industry will generally want to maintain their influence over the sector they service and may be closely allied with the key interest groups of their constituency (*e.g.* agriculture with farmers; trade and industry with the private sector). **Agencies responsible for natural resource management** (*e.g.* forestry, fishery) may face conflicting demands between their role to promote production and the need to use the resource sustainably over the long term. **Environmental agencies**, generally the newest and weakest agencies, often focus on regulatory approaches to environmental problems and may lack the economic skills to engage in debates over EFR. Hence, they may also not be particularly strong promoters of EFR, unless some of the revenue mobilised is allocated to allow them to perform their regulatory functions.

d) Sub-national entities

Some environmental problems are highly local or region-specific, while others spill over regional or even national boundaries. Environmental management must take account of these differences, leaving room for authorities at various levels to set or enforce standards in line with local circumstances. Decentralisation and the devolution of government functions to sub-national levels – whether provincial, state or more local level – can provide important opportunities but also raise special challenges. These relate, in particular, to the distribution of the benefits from the exploitation of natural resources such as forests and, conversely, the distribution of the cost of protecting natural resources, such as rivers, which cut across administrative boundaries. They also relate to the respective rights and responsibilities of national and sub-national entities with respect to taxation and regulation. Sub-national entities generally have a strong interest in raising funds, to reduce reliance on inadequate or unreliable transfers from central government.

e) Political leaders[4]

Political leaders are often closely linked with specific constituencies and sometimes have sufficient influence to resist reform proposals which would harm those constituencies. On the other hand, newly elected or particularly forceful political leaders may want to signal to their electorate that corruption and protection of special interests will not influence policies and may be willing to champion controversial forms of EFR.

f) The media and civil society organisations

The media, and civil society groups, such as religious groups, trade unions and professional groups, play a key role in presenting reforms to the public and creating environmental awareness. International environmental NGOs also exert pressure via the international arena. Their perspectives will largely depend on their particular political or ideological backgrounds, and on the scope for voicing independent points of view. In many countries, the government will have a strong control over the media.

g) Donors

Development agencies play important roles in areas such as public expenditure management and processes related to poverty reduction strategies. They generally are seen, including by themselves, as neutral providers of advice, based on "international best practice". However, they also have their own institutional biases, priorities and obligations, *e.g.* regarding accountability to their governing bodies. In some cases, donors' policies may be influenced by other objectives, ranging from facilitating access to certain sectors by their private sector investors to the promotion of exports. (See the chapter below on "policy coherence" for development.) In some cases, the visibility and influence of donors can affect the way their policy recommendations are perceived. For example, advice from the Bretton-Woods Institutions is often seen as key by the ministry of finance, while NGOs may dispute it precisely because it originates from them. In many cases, it is convenient for the government to be able to "blame" an external agency for "forcing" it to impose an unpopular policy which it would have intended to pursue anyway.

Allocating revenue from EFR

How the revenue raised (or freed up) as a result of a reform is allocated and used is a critical component of the reform design and implementation, and it will strongly influence the perceived and actual impact on affected stakeholders. Revenues from EFR can be affected to general or pro-poor expenditures, used to compensate those most negatively affected by the reform, and used to strength the capacity of concerned agencies. In some cases revenue has to be used to reduce opposition to reform originating from non-poor but politically influential groups. The pros and cons of various approaches are reviewed below.

Channelling revenue to the general budget or for general poverty reduction-related expenditures, such as education, which may have no direct link with the environment. This option, generally preferred by the finance ministries, ensures the flexibility of government spending and the rigour of the budget allocation system. It can help alleviate a budget deficit, reduce public debt, or reduce other distortionary taxes.[5] Another important consideration is the objective of integrating EFR into national planning and budgeting processes, notably in relation to comprehensive policy frameworks such as poverty reduction strategies. On the other hand, channelling revenue to the general budget may be relatively unpopular insofar as the benefits from the reforms become abstract or diffuse, such as when, for example, the additional revenue is used for "invisible purposes" such as to reducing public debt. Many people also feel that "environmental" taxes should at least partly be used for environmental purposes.

Revenue earmarking. The proceeds from taxes can also be allocated to pre-determined uses (i.e. "earmarked"). These include infrastructure investment, social measures or monitoring and enforcement activities connected to the environmental measure itself.[6] In many OECD countries taxes on motor fuels are often used to finance road infrastructure.

Environmental authorities have traditionally advocated earmarking of revenue from environmentally-related charges for financing environmental projects either through general budgets or through public environmental funds (budgetary or off-budgetary), controlled by the ministries of environment. This has often driven environmental authorities into conflicts with fiscal authorities and their foreign advisers (such as IMF). It is widely acknowledged that earmarking introduces rigidities in the tax system and obstacles to the continuous re-evaluation and modification of the tax and spending programs. It may also encourage a focus on the revenue potential of taxes to the detriment of environmental objectives. In addition, it creates precedence and gives rise to the claims of other government agencies (e.g. those which deal with forests, water, agriculture, education, etc.) for the right to have their own earmarked funds. This may lead to a budget fragmentation, making the economy impossible to manage. Earmarking rules, once instituted, are often difficult to reverse or revise. For instance, allocating fuel taxes to road infrastructure may lead to over-investment relative to priority needs in the same or other sectors. This problem is significantly reduced if revenues are earmarked for sufficiently broadly defined objectives, although the credibility of earmarking may also be reduced. In some countries, earmarking is prohibited by law.

Partial earmarking. Notwithstanding the limitations noted above, at least partial earmarking may, as a second best, be a price worth paying for having predictable financing for environmental monitoring and enforcement activities, notably in countries where environmental agencies are poorly resourced or heavily dependent on donor assistance. It may also reinforce incentives to enforce environmental regulations and collect associated taxes and fees as well as help generate public support for EFR.

In such cases, the rationale for earmarking should be evaluated rigorously and on a regular basis to avoid inefficient spending. Similarly, environmental funds established with earmarked funds must be managed in line with internationally-recognised principles of sound public expenditure management, relating to fiscal discipline, efficient allocation of public funds, operational efficiency, accountability, transparency and comprehensiveness of the budget. *Good Practices of Public Environmental Expenditure Management in Transition Economies* (OECD, 2003) provide specific guidance in these regard. These include checklists for assessing performance with respect to environmental effectiveness, fiscal prudence and management efficiency.

Financing transition costs. Revenue from EFR earmarked revenues can be channeled back or "recycled" to the industries most affected by the taxes to help them adjust. Such temporary assistance can be a way to address concerns that taxes will affect competitiveness relative to firms in other countries who do not face similar taxes.

In order to avoid effectively subsidising continued pollution, this must target those firms which take steps to address pollution, and not simply return to each polluter the taxes paid. Revenue neutrality for an industry as a whole does not necessarily mean revenue-neutrality for individual businesses, some firms will lose and some firms will gain, depending on the design of the tax and the recycling mechanism. Recycled revenues can also be allocated to collective environmental infrastructure (*e.g.* waste water treatment plants) or to fund "generic" research and development which benefit the whole industry.

Revenue recycling is an effective way to increase the acceptability of reform by industries. In China, for example, a combination of pollution taxes and support for pollution abatement expenditures has proved effective. Effective use of recycled funds requires a system for rigorous cost-benefit assessment, to ensure that the most cost-effective pollution control investments receive support. Revenue recycling aimed at facilitating adjustment should also be *strictly time-bound*.

Compensatory cross subsidies. Differential pricing for different categories of users (generally charging heavy users more) can be a way to mobilise funds to compensate for the negative impacts of the introduction of pricing for water or electricity on poor households. As for all subsidies, these must be carefully designed and targeted, to actually benefit the poor

ENVIRONMENTAL FISCAL REFORM FOR POVERTY REDUCTION – ISBN 92-64-00868-3 – © OECD 2005

without undermining incentives for conservation. They should also be transparent and limited in time to avoid negative implications associated with subsidisation (IEA/UNEP, 2002).

Sharing EFR revenues between different levels of government. How revenues are shared between different levels of government raises critical issues. Allocating revenues from pollution or natural resource extraction taxes to the regional or local government of the region where they are collected can stimulate collection efforts and public support for the taxes. On the other hand, equity and other concerns often call for reallocation of resources across different regions of a country, through the national budget. In the case of natural resource taxes, there can be tensions between the need for local government and communities to benefit from the taxes and the need to re-allocate resources towards poorer regions. Where revenues from EFR are significant – the appropriate division is often a source of much debate and conflict between the centre and states. Issues of capacity also arise: in many countries, even decentralised ones, subregional authorities lack the administrative capacity to collect and manage tax revenues.

The EFR policy cycle and the role of various stakeholders

A stylised "policy cycle" would involve a sequence of "agenda setting"; assessment of policy options, decision-making and implementation followed by monitoring and revision as needed. Real-world policy development and implementation seldom follow this neat pattern, and the various phases of the cycle are inter-linked and overlapping. It is useful, however, from an analytical perspective, to distinguish between different phases of the cycle and the respective roles of different stakeholders during each phase.

Figure 3.1. **Stylised representation of the stages of the EFR "policy cycle"**

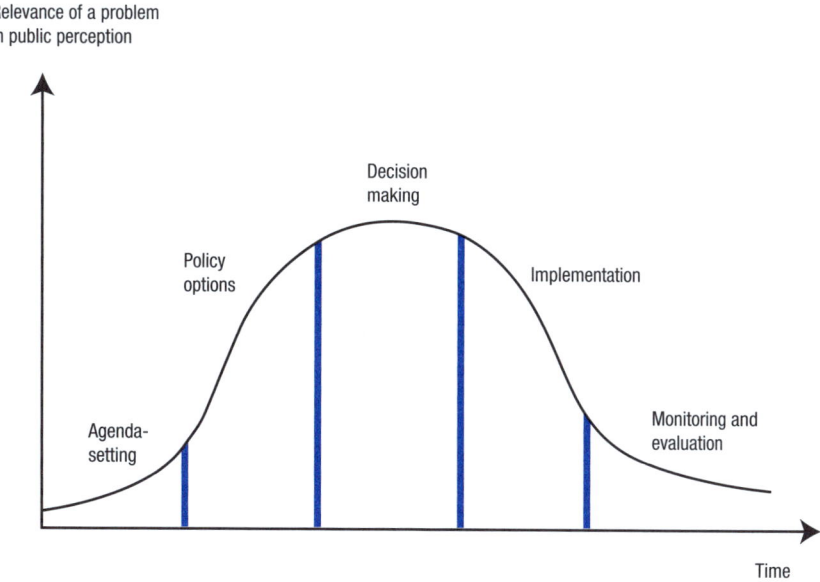

Key steps in EFR formulation and implementation include:

Initial research and agenda-setting. EFR starts with a sound understanding of a given problem (for example, the health impacts and associated costs of pollution) and possible remedies. This usually involves extensive research and includes an assessment of various instruments which could be used (including fiscal and non-fiscal) for achieving the objectives most efficiently. The links with social and economic objectives are also assessed. Having a strong analytical basis readily available, and making it publicly available, is essential to be able to seize on opportunities for reform arising, for example, from unexpected political or other events.

Policy development. This generally involves thorough consultation with stakeholders, in order to further substantiate and refine the results from the initial research. Once decisions have been made on the type of policy mechanism to use, there is further consultation on the detailed design of the overall package, including how to collect and allocate revenue from environmental taxes. This must be considered within the context of existing policies and institutions, to ensure a match between the instruments proposed and their feasibility and cost-effectiveness. EFR proposals often combine fiscal, regulatory and voluntary approaches and allow for measures to compensate losers, and to encourage compliance. Efforts to combat industrial pollution may combine price and regulatory measures, public information campaigns ("shaming the polluters") as well technical assistance for affected industries. Such combinations of instruments help achieve "win-win" solutions by counterbalancing or compensating for some of the negative effects, and helping mobilise support for the reforms.

Dialogue, information dissemination and advocacy. Identifying a problem and possible solutions is not enough. Acceptance of EFR proposals is an issue of marketing. EFR – just like other reforms – may have to be literally "sold" to the public through public awareness campaigns. This requires accurate information presented in a way that is easy to understand. Dialogue between key stakeholders will also generate additional information on stakeholders' perspectives. Consultations and dialogues with key stakeholders in the policy development process (outlined above) will also help generate support for the policy. Dialogue can also help to form political alliances and get political majorities where legislation is required. However, it should not be assumed that dialogue will lead to consensus, and choices between different interests will often remain. Where the problems associated with corruption and patronage are serious, resistance to EFR and particularly around natural resources will be particularly strong and building strong alliances is vital. Concerted efforts of fiscal and environmental agencies with inputs from civil society, including research and policy institutions are often necessary. For example, in Cameroon, recent studies showing lost forest revenues of over USD 100 million per year from illegal logging and low rent collection helped the government – especially the ministry of finance – to rally support for reforms.

Advance notice and gradual implementation. Public announcement, preferably early on, is important to give affected parties the time to prepare and adapt to the proposed changes. Gradual phasing-in of reforms is another way to reduce the transition costs. It may also help build political support for keeping the reform process on track as firms which have undertaken investments in anticipation of planned reforms will want to see those investments pay off. Step-wise implementation also allows to test new instruments and build up experience and confidence.

Monitoring and evaluation. This is essential both to identify and remedy implementation problems as they emerge as well as to verify the benefits of the intervention. Good monitoring and evaluation may help identify unexpected and unintended consequences of the reforms, calling for revisions in approaches and objectives. Publishing widely the amount of revenues raised by EFR and how they are spent can enhance accountability and public support. Lack of monitoring opens up an accountability gap and leaves more space for corruption.

Building the credibility of monitoring and enforcement agencies. Credibility rests on transparency and public participation. Transparency helps reduce corruption, while public participation helps mobilise and sustain support for reform. Credibility is essential to sustain support for reform, and confront challenges from, for example, affected industries which have direct interests in portraying the monitoring agency as unprofessional, corrupt, or abusive.

Different stakeholders play a leading role in the course of the policy cycle:

Academic institutions have generally played a pioneering role with regard to developing initial EFR concepts, and agenda-setting, encouraging rigorous assessment of various options and approaches and encouraging wider discussions via **the media**.

Civil society groups often play a lead role analysing the impacts of EFR in their respective sectors and making them understood in their respective constituencies. Thus, NGOs often contribute to raising awareness of environmental and health problems amongst the wider public and enhancing political commitment for EFR. In Germany, for example, some NGOs specifically dedicated to promoting EFR emerged.[7]

The general public. Awareness of environmental issues, which rose during the 1970s and 1980s, has been a critical factor underpinning EFR initiatives in most countries. The idea that the tax system should move away from taxing "goods" to taxing "bads", such as pollution, is often well received by the public as a "win-win" proposition, increasing political support for reforms. **The Media** often play a critical role in fostering this awareness.

Political actors play a critical role in helping translate academic proposals into politically viable legal initiatives balancing political positions, and ultimately enabling the enactment of reforms.

Governments have the lead role in formulating priorities, ensuring that EFR proposals are consistent with existing tax and budget systems, assessing the merit of various options and instruments and their administrative feasibility, and developing and implementing policy reforms.

Donors can play a supporting role in each of the phases of the "policy cycle". This is taken up in the next chapter.

Notes

1. This is by no means unique to EFR. The same applies to many policies – such as structural adjustment, privatisation of major enterprises or trade liberalisation.

2. "Interests" are used as each group is not just a passive recipient of reforms, but will often actively engage – to initiative, support, or undermine – aspects of the reform process.

3. The term "government bureaucracy" is used here to refer to the government ministries involved in the reform process.

4. The term "political leaders" is used here to refer to the heads of the executive (*e.g.* President), members of parliament, political parties, and other key political leaders.

5. In OECD countries, EFR has often been undertaken in the context of a constant tax burden, and governments that have introduced environmentally-related taxes have reduced distortionary taxes, in particular through cutting employers' social security contributions.

6. This does not apply to user charges which are by definition always "earmarked" to cover the costs of corresponding service provisions.

7. The most famous among them was the Förderverein Ökologische Steuerreform ("Society for the Promotion of Environmental Tax Reform").

ENVIRONMENTAL FISCAL REFORM FOR POVERTY REDUCTION – ISBN 92-64-00868-3 – © OECD 2005

ISBN 92-64-00868-3
Environmental Fiscal Reform for Poverty Reduction
© OECD 2005

PART I

Chapter 4

Conclusions and Recommendations for Donors

Basic facts about EFR

"Environmental fiscal reforms" include a range of taxation and pricing measures which can raise fiscal revenues while furthering environmental goals. EFR approaches alone will not solve countries' fiscal and environmental challenges but they are an important part of the policy tool kit.

EFR can contribute to poverty reduction *directly* by helping address environmental problems – such as water contamination and air pollution – that impact the poor. EFR can also help *indirectly*, by generating or freeing up resources for investment in infrastructure which is critical for the poor, such as water supply and sanitation. EFR can also help free-up budget resources for other kinds of pro-poor investments, such as health and education. EFR can therefore play an important role in pursuing the Millennium Development Goals of eradicating extreme poverty and hunger and of ensuring environmental sustainability.

The range of EFR instruments is wide. Instruments relevant to developing countries include taxes on natural resource extraction such as fisheries and forests, cost recovery and pricing measures to improve access to basic services such as water and energy, taxes and subsidy reforms to discourage the use of environmentally damaging products, as well as taxes and fees to control harmful industrial pollution and waste. Different EFR approaches and instruments are applicable to different sectors and issues. The main instruments of EFR are examined in Chapter 2 above.

EFR instruments do not substitute for but complement and strengthen regulatory and other approaches to fiscal and environmental management. EFR instruments should therefore be thought of as components of fiscal and environmental policy mixes, not as "stand alone" policy packages. EFR must be viewed in the context of comprehensive policy mixes combining fiscal, regulatory and other instruments to achieve sound fiscal and environmental management.

Comprehensive approaches to development, such as "poverty reduction strategies" and "sector-wide approaches" provide important opportunities for EFR. These approaches provide new opportunities to integrate EFR in the context of country-led national and sectoral development plans. Medium-term expenditure frameworks (MTEF), in particular, address directly issues relating to fiscal revenue collection and tax and pricing reforms.

EFR challenges and instruments vary widely across different economic sectors. Issues related to natural resources extraction, industrial pollution, water supply and energy call for different approaches and instruments. In Part II of this document, below, EFR is examined in relation to specific economic sectors of particular relevance for developing countries.

But issues of "good governance" always assume a central role. In this respect, EFR is similar to other forms of economic policy reform. The design and implementation of EFR entails a number of trade-offs. Political and institutional issues, including transparency and participation in decision making, thus play a central role.

Key challenges of EFR design and implementation

The success of EFR implementation depends on the interaction between a range of factors. These include: the characteristics of the problems addressed and its perceived priority; the general social, economic and political context; and opportunistic factors such as the election of a popular new government, or less positively, an environmental disaster.

EFR design and implementation must take prominent account of political feasibility, in addition to equity, fiscal and environmental effectiveness and administrative feasibility. Experience in many countries has shown that planning for adoption and implementation of EFR measures must take account of likely political resistance. The "political economy" dimension of EFR is key.

Key steps include identifying likely winners and losers, and understanding the perspectives and interests of affected stakeholders, such as political leaders; poor and vulnerable groups; the private sector, the government; civil society groups; and the media.

How the revenue raised (or freed up) as a result of a reform is allocated and used is a critical component of policy design and implementation. It will strongly influence the perceived and actual impact on affected stakeholders. These issues are examined in detail in Chapter 3 above.

The EFR policy cycle and donors' roles

The EFR policy cycle involves a number of phases. These include initial research and "agenda-setting"; "policy development"; "dialogue, information dissemination and advocacy"; "advance notice and gradual implementation"; "building the credibility of monitoring and enforcement agencies" and "monitoring and evaluation". These phases are related, often overlapping and iterative over the EFR policy cycle.

Donors can play important roles at each stage of the EFR policy cycle:

a) Agenda setting stage

EFR must start with a sound understanding of the issue to be tackled, including notably its impact and causes and its relative importance in view of the many pressing issues facing a country. This analysis must be based on a sound scientific basis. This includes identifying the social and economic consequences of environmental problems and comparing environmental quality standards and health status in a given country with performance in comparable countries. Solid information is critical to raise the attention of policy makers and the public as to the severity of a given issue, to challenge adverse perceptions and overcome opposition from vested interests. The media often plays an important role in disseminating this information and raising public awareness.

Donors can play an important role in this regard by supporting the work of universities, research groups and international organisations in relevant areas.

b) Policy development stage

"Policy Development" involves an assessment of the mix of instrument (fiscal and non-fiscal) which could be used to address the problem identified most efficiently, given existing socio-political and institutional conditions. This must in particular establish that EFR measures are the most suitable for the problem at hand. It also involves quantifying the expected fiscal, environmental and social benefits. It also includes identifying potential "winners" and "losers" from reform and formulating possible compensatory measures. Having a solid analysis of a problem and well-developed policy approach to deal with it is essential to be able to seize opportunities for reform arising, for example, from unexpected political or other events.

EFR policy packages often combine economic, legal and voluntary approaches. Detailed policy design must ensure that the new policies and instruments proposed are feasible and cost-effective and supportive of the overall policies of the government and any other planned reforms. This includes an assessment of long-term social, economic and environmental impacts. It is also essential to ensure coherent policies across sectors and to maintain the integrity of the national budget system. Examination of how revenue from EFR can be collected and allocated must therefore take account of the entire fiscal system of the country.

Policy development is an inter-ministerial process involving all relevant ministries and agencies. It also involves broad-based consultation with affected stakeholders (including representatives of civil society, the private sector and vulnerable groups) in order to improve understanding of their perspectives of how adjustment can be facilitated, and how to minimise the costs of compliance. Dialogue with stakeholders can help build support for reform. Dialogue can also help to form political alliances and get political majorities where legislation is required. This is often the stage at which coalitions of different stakeholders in support of reform can emerge.

Through their support to various economic sectors, notably through poverty reduction strategies and sector-wide approaches, **donors** can encourage integrated, cross-sectoral policy reforms including in relation to EFR. This includes, in particular, ensuring that available opportunities for "win-win" EFR approaches are not missed. Donors can also support those sectors of governments (such as finance or environment ministries) who favour reform to overcome bureaucratic inertia or resistance from hostile ministries or agencies. Donors can encourage transparency, access to information concerning public finances, public participation, and accountability, which are key preconditions for sound policy development and, more generally "good governance". They can support capacity development, notably in relation to the measurement of implicit taxes and subsidies and the quantification of impacts. They can also support research groups and universities, NGOs and the media to participate in the assessment of proposed reforms, notably in relation to their impacts on disadvantaged groups and on the environment.

c) Dialogue, information dissemination and advocacy stage

Identifying a problem and possible solutions is not enough. Securing and political acceptance and public support for EFR proposals often requires active advocacy, including through public awareness campaigns. This requires accurate information presented in a way that is easy to understand. However, it should not be assumed that dialogue will lead to consensus, and choices between different interests will often remain. Where the problems associated with corruption and patronage are serious, resistance to EFR and particularly around natural resources will be particularly strong and building strong alliances is vital.

Concerted efforts of fiscal and environmental agency with inputs from research and policy institution are often necessary. In Cameroon, for example, coalitions for reform in the forestry sector involved environmental agencies and NGOs as well as the ministry of finance – to rally support for reforms.

Donors can contribute to such dialogues and awareness-raising, including through support to civil society groups. They can also support international sharing of experience and dialogue on EFR among developing countries governments, international organisations and NGOs. Examples of this include current initiatives, for example, on the transparency of extractive industries (Extractive Industries Transparency Initiative – EITI) and illegal logging (Forest Law Enforcement Governance and Trade Process – FLEGT). Donor agencies can also make available information on reform experiences from OECD countries, bearing in mind that many OECD countries also face various challenges in implementing successful EFR.*

d) Implementation stage

Implementation generally starts with public announcement of upcoming reforms. This should preferably be done early on as it is important to give affected parties the time to prepare and adapt to the proposed changes. Gradual phasing-in of reforms is another way to reduce the transition costs. It can help build political support for keeping the reform process on track as firms which have undertaken investments in anticipation of planned reforms will want to see those investments pay off. Step-wise implementation helps test new instruments and build up experience and confidence.

Donors can play an important role by helping financing the transitional costs of reform, in order to protect the poor from negative impacts or to overcome politically powerful blockages. This also includes supporting technical co-operation to help industries adjust to change (*e.g.* by switching to cleaner production techniques).

e) Monitoring and evaluation stage

This is essential both to identify and remedy implementation problems as they emerge as well as to verify the benefits of the intervention. Monitoring and evaluation may help identify unexpected and unintended consequences of the reforms, calling for revisions in approaches and objectives. Publishing widely the amount of revenues raised by EFR and how they are spent can enhance accountability and public support.

There is an important role for non-governmental actors to play in monitoring of the implementation and impacts of specific reform measures, and even in the actual collection of, for example, natural resource taxes. Particularly with weak and under-financed decentralised administrative structures, civil society groups can provide important complementary "policing" services that feed into ensuring implementation and follow-up of EFR legislation and regulation.

---

* For instance, while various OECD countries have substantially extended the use of environmental taxes since the early 1990s, there are currently still many exemptions and tax rate reductions granted to sectors most exposed to international competition.

f) Building the credibility of monitoring and enforcement agencies

Credibility rests on transparency and public participation. Transparency helps reduce corruption, while public participation helps mobilise and sustain support for reform. Credibility is essential to sustain support for reform, and confront challenges from, for example, affected industries which have direct interests in portraying the monitoring agency as unprofessional, corrupt, or abusive. Environmental agencies must also be credible *vis-à-vis* ministries of finance. This includes, in particular, adhering to existing rules and principles of public expenditure management. This is particularly crucial when environmental agencies are entrusted with the collection and management of taxes or fees or when the proceeds from environmental taxes and fees are earmarked and transferred to them for environmental management purposes. The capacity of fiscal authorities to collect and administer existing taxes (including compliance monitoring and enforcement) may also need to be enhanced for proposals for fiscal reforms to be credible.

Donors can play important roles in providing technical assistance to develop the capacity of monitoring agencies. In the case of natural resource, this may include the formulation of new management regimes as well as the formalisation and registration of property rights. It also includes building up the capacity of environmental agencies with respect to, for example, budgeting and public expenditure management. Fiscal authorities may also need to strengthen their capacity in various domains.

General principles to guide donors' actions

In their actions in support of EFR, donors must:

a) Emphasise country ownership and be sensitive to the local context

This implies taking full take account of local realities and conditions and embed EFR in political processes. There must be in-country demand for EFR. Donors should encourage country ownership and not force the pace: there must be a "home-grown" demand for reform. Strong country ownership, in turn, encourages donor harmonisation and will shield countries from excessive donor influence. Donor agencies also have to be sensitive to political challenges of EFR and attuned to the cultural and historic particularities. They should avoid imposing "blueprints", but instead provide financial, technical, institutional and political support to countries' own efforts.

b) Be prepared to act opportunistically

In a politically and economically volatile environment, it is crucial to take advantages of windows of opportunities when they arise. A new political leadership – especially if it has popular support – can be the catalyst for major policy shifts. Donors should be prepared to seize on the new opportunities for reform this creates.

c) Be pragmatic

It may be necessary and justified, on some occasions, to deviate from standard fiscal practice in order to secure political will and/or public support for important reforms. At least partial earmarking – assigning tax revenues from EFR to a particular use may in some cases be necessary to achieve progress and should be considered, given adequate safeguards. In such cases, donors can play an important role in working with fiscal authorities to ensure that appropriate safeguards are put in place to ensure consistency with efforts towards fiscal consolidation.

ENVIRONMENTAL FISCAL REFORM FOR POVERTY REDUCTION – ISBN 92-64-00868-3 – © OECD 2005

d) Strive towards harmonisation, alignment and policy coherence

In EFR, as in other areas, donor support should be harmonised – with donors working together to avoid duplication of missions, studies or institutions. And it should be aligned behind country-owned and led strategies – and procedures. Accordingly, EFR measures should be conceived within the context of PRSPs, sustainable development strategies, medium-term expenditures programmes and sector-wide approaches. More generally, donors should also ensure that their efforts toward poverty reduction, environmental sustainability and sound economic and fiscal management are mutually supportive.

Coherence of donor country policies on environment and EFR is also vital: donor countries can undermine the objectives of their development co-operation policies by inconsistent policies in other areas. Donor agencies should strive to raise this issue within their governments to draw attention to this risk and try minimise it. For example, donor agencies should raise awareness in their governments of agricultural policies which distort world market prices or fisheries agreements with developing countries which run against environment and poverty reduction objectives.

PART II

Environmental Fiscal Reforms: Political Challenges and Opportunities in Selected Sectors

This part reviews the political economy challenges of EFR in the areas of forestry; fishery, energy, transport, water; and industrial pollution control.

ISBN 92-64-00868-3
Environmental Fiscal Reform for Poverty Reduction
© OECD 2005

PART II

Chapter 5

EFR in the Commercial-scale Forestry Sector

Research, revise and amend policies on standard price norms in caring for protected forests, afforestation and protective forest management. (Vietnam Poverty Reduction Strategy, 2003)

Key features of the sector: overview

Natural forests are an important resource in many developing countries. Apart from timber, they provide goods such as firewood and a wide range of non timber forest products (*e.g.* wild fruits and roots, grasses, vines, mushrooms, medicinal substances, gums, honey, game, meat, etc.) which are critical to the livelihood of the poor, notably the landless.

Forests also provide a wide range of services such as watershed preservation and the regulation of water cycles, thus playing a major role in protecting surrounding and downstream lands, and in preventing the degradation of water supply, hydropower, irrigation, transport and other important man-made infrastructure. In addition, forests provide habitats for a wide range of wild species, play a major role in carbon sequestration and embody important cultural and spiritual values for many communities. The "public good" nature of natural forests (as distinct from tree plantations) is reflected in the fact that they are formally under state ownership in most countries.

Timber, the most visible of all forest products, provides the raw material for many industries, including for export, generating significant employment and fiscal revenue in many countries, particularly in south-east Asia, west and central Africa and Latin America. However, timber extraction is taking place at an unsustainable pace, and is an important cause of forest destruction. Moreover, the profits generated by this destruction are mostly captured by the private sector, with limited benefits for society at large. Governments in countries such as Brazil and Indonesia capture less than 15% of potential rent, while this percentage reaches around 30% in Gabon and Laos.

Improving forest management regimes is a priority both for the purposes of improving revenue capture by the state and for ensuring long term sustainability of this important economic activity. Fiscal instruments can play an important supporting role in this respect, complementing regulatory approaches. We focus here on policies and instruments concerning commercial scale forest exploitation by private sector operators. These are most relevant for forest-rich countries where commercial-scale forestry is a significant economic sector. Community-scale operations are not covered as they raise significantly different issues.

Note that the discussion below applies to *natural forests* as distinct from tree plantations (which generally involve a limited number of fast-growing species), which have limited non-market, environmental and non-timber benefits.

National Forest Programmes (NFPs) can provide a good entry point for EFR in the forestry sector. They involve multi-stakeholder dialogue processes regarding all forest-related issues, and provide a framework for developing adapted financing strategies and policies for the sector. Many donors provide support for NFP even though their respective programmes may not bear that particular label in all countries.

Although this chapter focuses on commercial-scale timber extraction, other goods and services play an important role in sustainable forest management. While most public financial revenues from forests are linked to timber extraction, the many other forest products and services, also have economic value, and contribute to welfare. The valuation of these non-marketed products and services and the various institutional, policy and governance issues affecting their sustainability are important issue taken up in NFP processes. These issues lie beyond the scope of this chapter.

Policies and instruments

Given appropriate policies, regulations, technologies and management systems, timber extraction, revenue generation and the long-term sustainability of forests – and corresponding non-timber goods and services – can be compatible. The main challenge of forest policy is to reconcile the multiple functions of forests. A first step in this respect is to distinguish between forests which can be used for timber extraction purposes, and those which are too fragile, depleted or otherwise degraded; and to recognise those on which many landless poor or indigenous communities depend as well as those with high cultural, social and spiritual value and which should be preserved from commercial-scale or even any logging activities.[1]

The choice of appropriate approaches and instruments for forest management is highly dependent on country-specific factors relating to legal and regulatory frameworks, institutional capacity, ecology, topography and others. It is very difficult to make generalisations. There is no "one size fits all" model. This chapter thus provides an overview of the pros and cons of various approaches and key issues to consider.

a) Regulatory instruments

Within the "commercial forests" estate, ensuring sustainable logging implies imposing a set of rules on private sector logging firms. Thus, leases over timber lands prescribe selective logging and a variety of qualitative and quantitative regulations aimed at preserving the long-term productivity of the forest. Given that forest leases tend to be "short term" – in relation to the growing period of trees, which spans over decades – the incentives to restrain short term profits for the sake of long-term sustainability are very weak. Enforcing regulations which reduce profits for loggers is a critical challenge of forest management.

b) Fiscal instruments

There are several ways to capture rents from timber extraction. They include "stumpage taxes" levied on timber harvested – either by value or volume – or on timber exported; charges per hectare of concession, taxes on corporate profits or income taxes; state participation in the industry; and auctions of timber concessions combined with refundable performance guarantee deposit systems. These instruments are often used in combination.

c) Setting the level of the tax

When set at an appropriate level, taxes on logging could i) reduce incentives to obtain logging concessions (by reducing profits); ii) generate financial revenues for the general budget, thereby raising the profile and attention paid to sound forest management by the government; and iii) generate revenue to strengthen indispensable monitoring and enforcement efforts (partial earmarking of the proceeds from forest taxes).

However, estimating the true level of economic rent from timber extraction, and thus estimating the right amount of tax to collect, is not straightforward for three reasons:

1. The cost of forest management (which includes the cost of logging and the costs of managing the entire concession, including immature stocks) varies according to the degree to which the operator abides by the logging rules. Thus a logger abiding by the rules will face higher costs.[2]

2. Costs of sustainable logging are highly location-specific. A tax set at an equal level for all runs the risk of penalising those following good practices, including those operating in more fragile environments, which requires more careful and costly logging techniques.

3. The costs of sustainable logging include a significant share of fixed costs which do not fluctuate much over time, while export prices of logs can fluctuate significantly. A decline in export prices and therefore revenues cannot be matched by a proportional reduction in costs. Even if no logging takes place in a given period, the forest has to be managed and the immature stock maintained. Unless monitoring and enforcement systems are very rigorous, overtaxing creates incentives for unsustainable logging. Regardless of the system used, a key requirement is therefore to avoid overestimating the actual profits generated by forest exploitation and not "over-tax". "Ecolabelling" approaches, which aim to ensure that sustainably extracted timber receives a premium in the market, reflect this concern.

d) Choice of instruments

Stumpage taxes are very widespread, but also the most open to abuse through evasion or corruption. As they require log measurement in the forest, such levies encourage corruption of those charged with measuring, leading to under-reporting of harvests. Forestry staff, who often depend on the concessionaire for transport and accommodation, and whose salaries are tiny relative to potential gains from violations of forest rules, are obviously vulnerable to persuasion, pressure or bribery. These considerations as well as administrative simplicity and cost efficiency considerations suggest that it is preferable to collect timber taxes at the point where they are loaded onto a boat or at the gate of the processing plant, rather than in the forest. This has the additional benefit of keeping distinct the functions of revenue collection from those of monitoring and enforcing logging rules, thereby reducing the scope for corruption.

Log export taxes have the advantage of administrative simplicity and ease of enforcement. However, by making domestic timber relatively cheaper though, they can discourage efficiency in domestic processing industries.

Excise taxes at "saw-mill gate". Where domestic use is significant, excise taxes could be collected at the level of the processing plant. If saw mills are numerous and small scale, however, this may be very difficult to administer.

Concession fees include one-time licence fees upon concession award and annual fees based on concession size. Provided they are set and levied in a transparent and open manner, they can help reduce dependence on volume-based fees which are difficult to collect and easy to evade. They also have the advantage of discouraging the acquisition of large concessions. To date, they are not widely used and generate little revenue.[3] In many countries, the award of concession is far from transparent, often leading to corruption on a large scale.

Auction mechanism. Concessions can be sold through a bidding process, thus allocating the concession to the party to pay the most to secure access.[4] This raises revenue and provides a market-based indication of the *financial* value of the concession and raises revenue. If used in conjunction, auction mechanisms and forest fees are closely linked: if forest fees are relatively low, operators will be willing to pay high bids. Conversely forest fees which capture the full *financial value* of the timber will result in low bids. Success of auctioning mechanism requires a competitive environment to prevent collusion between larger timber firms. Auction mechanism are best used in combination with performance guarantee bonds (see below).

Corporate income taxation. If the risks of corruption or of over-taxing is considered high, for example because the cost of sustainable logging are much higher than the costs of unsustainable logging, or because cost structures are very different across geographical areas, it may preferable to tax the profits of logging firms indirectly through corporate taxes rather than the timber volume of value.

By themselves, none of the mechanisms outlined above discourage non-compliance with logging rules. In addition to monitoring and enforcement, specific mechanisms can be used for that.

Performance guarantee mechanisms can provide direct incentives to comply with forest rules and other environmental regulations. The underlying principle is that concessionaires will post a bond in advance of harvesting which will be refunded pending verification of compliance with the stipulations of the concession lease. The value of the guarantee bond can be set by auction. Such systems are strongly complementary with auction mechanism as well as eco-labelling approaches, which attempt to ensure that sustainably harvested timber obtains a premium on the market. However, few countries have tried this approach to date.

Perspectives and interests of affected stakeholders

Forest-dwelling communities. These include ethnic minorities and indigenous communities who are often amongst the poorest. They are often displaced following the formal award of the forest in which they live to commercial operators. In some countries, community forests are encroached upon or illegally logged. Forest-dwelling communities stand to benefit directly from more transparent policies and rigorously enforced forest policies, and the containment of patronage-based concession award. The extent to which they benefit from increased revenue from forest exploitation depend on the share of such revenues allocated to the regions where they live rather than to the National Treasury. The employees of logging firms, who include local communities but also migrants, stand to gain from more professional management in the logging sector. However, those working for companies engaging in illegal practices stand to suffer from rigorous enforcement, at least in the short term.

Downstream communities. Downstream communities benefit of services provided by forests, especially of those protecting the watershed of erosion and provision of water. Farmers relying on irrigation stand to gain directly from improved watershed stability.

Political actors. In many countries, forest concessions are used by elite groups as sources of finance and influence. Many concessions are awarded to cronies with little or no attention paid to sustainability imperatives. In some cases, timber profits finance violent conflict. Political actors relying on forest as sources of funds have a strong interest in resisting reforms. Issues relating to the distribution of revenue from forest give rise to conflicts between national level and local-level political actors.

Interests of investors. Domestic and foreign investors in the harvesting industry range from capital-intensive, skilled and generally environmentally sound operators to small-scale "fly by night" operations. In some countries, such as Cameroon, large foreign investors dominate the industry, while in others, such as Indonesia, domestic firms are key players. Established companies – domestic or foreign – often have a strong interest in undermining reforms which would increase competition and reduce their profitability. On the other hand, companies which are excluded from the industry due to lack of political connection often welcome reform. Domestic timber processing industries generally favours log export bans or taxes as they artificially lower domestic log prices below world market levels.

One critical concern of large-scale investors is the stability of the concession regime over time. Unpredictable and frequently changing regimes (including changes in taxation levels, introduction of new charges) act as powerful disincentives to long-term investment in sustainable forestry. In an industry where rotation periods span over decades, the risk of not being able to profit from an investment made when it finally bears fruit is extremely high. Reforms which increase the stability and predictability of regulations are thus welcome.

Ministries of forestry face huge demands of regulating and monitoring a geographically dispersed resource with limited staff and transport support – and *policing* a considerably better resourced private timber industry. They have to handle the conflicting tasks of protecting the resource base, while, on the other hand increasing timber outputs to feed downstream industries. They often share government-wide weaknesses of low pay, lack of accountability and hence corruption. These ministries stand to gain from EFR, provided that an adequate share of the proceeds are earmarked for monitoring and enforcement activities. However, ministry staff relying on side payments to supplement their incomes have strong incentives to resist reforms.

Finance ministries generally focus on the potential revenue to be obtained from the forests sector and have to face issues relating to the distribution of these revenues between national and local-level authorities. However, they are often unfamiliar with the complex issues involved in fostering sustainable forestry and the non-market benefits of forests at the local or national level.

Development agencies and the international community. They have often played important roles in encouraging forestry reform processes. Reform has been promoted both by the carrot of projects to stimulate change in the forestry sector, but also the stick of holding

ENVIRONMENTAL FISCAL REFORM FOR POVERTY REDUCTION – ISBN 92-64-00868-3 – © OECD 2005

up the payments of World Bank and IMF structural adjustment to governments. In some cases, donor countries have played an ambivalent role in the reform process, when this threatened the commercial interest of their citizens.

Civil society: international NGOs, but also domestic NGOs, have been powerful advocates for change both on governments, but also on development agencies to hold governments to account. In Cambodia, pressures from international NGOs have been instrumental in promoting reforms, and in getting development co-operation agencies involved in the area.

Managing the reform process: key steps

a) *Information collection and dissemination*

Collecting and disseminating reliable information on the social, environmental and economic damages from unsustainable forest management – as well as the foregone revenues – is a key step in raising awareness of key stakeholders and the public. In forest-rich countries with a long tradition of timber exports, pointing out that dwindling supplies are leading to a need to import timber from abroad can be a powerful argument.

b) *Coalition building*

Experience suggests that coalitions of NGOs (local and international) and local communities are essential to foster reform in the forestry sector, and to counter the power of vested interests. Development co-operation agencies can play an important role.

c) *Building a competent and credible forest management agency*

In order to overcome corruption and the influence of political and commercial vested interests, capacity development for building a competent and credible forest management agency is critical. This includes strengthening capacities in areas such as design and negotiation of concession contracts; assessment of forest utilisation plans; quantification of the non-market values of forests; design of appropriate regulations and corresponding monitoring and enforcement approaches; on-the-ground monitoring and enforcement of those rules; and others.

Box 5.1. **Some guiding principles for forest policy**

There can be no "one size fits all" policy package for forest management policy. Policy mixes must reflect prevailing social, economic, political conditions and ecological conditions, as well as land tenure and property rights regimes, fiscal systems and institutional capacity. However some guiding principles can be identified:

- Enforcement of forestry rules cannot dispense with *on-site monitoring*, reinforced, if possible, by public oversight of forest authorities' performance. This monitoring must focus on the ecological state of the remaining forest, rather than on the flow of timber extracted. Public disclosure of violations can provide a powerful instrument.

- Focus on *preventing the degradation of immature stock*. This is more important for long-term forest management than reducing the profitability of timber harvesting.

- *Maximise security of tenure*. Rotation periods, often in excess of 30 years, combined with uncertainty about future timber prices, fluctuations in interest rates, and a host of other factors, make investments in long-term maintenance of forests very risky. Insecure tenure further discourages long-term oriented management.

- Provide *Means to Penalise Breaches of Rules While Economising on Monitoring*. As the owner of the forest, the government and the public bear the ultimate consequences of violation of logging rules. Tenurial instruments must provide the means to penalise such behavior.

- *De-link monitoring of forest regulations from tax collection activities*. This helps to avoid the inherent conflicts of interest involved and reduces the scope for corruption.

Source: Adapted from Leruth *et al.* (2001), "The Complier Pays Principle: The Limits of Fiscal Approaches Toward Sustainable Forest Management", IMF Staff Papers, Volume 48, Number 2, December 2001.

Notes

1. Countries which have signed the UN Convention on biodiversity must also take into account the protection of "biodiversity hot spots".

2. Certain techniques (for example reduced-impact logging) can minimise the negative impacts of logging and protect biodiversity. However, these techniques require high levels of know-how and skills. Also, given current timber prices, they are less profitable than the standard techniques, and firms will try to avoid them.

3. See, for example: Gray (2002), Forest Concession Policies and Revenue Systems. World Bank Technical Paper No. 522 (Forest Series).

4. See Gray (2002) for a discussion of the respective merits of "open" bidding, "sealed-tender" bidding, "oral" auctions and other auction mechanisms.

ISBN 92-64-00868-3
Environmental Fiscal Reform for Poverty Reduction
© OECD 2005

PART II

Chapter 6

EFR in the Commercial Fisheries Sector[1]

> *The primary objective (of the growth and macroeconomic stability policy) will be to ensure full mobilisation of domestic revenue… non-tax revenue will increase largely as a result of fishing fees and fines.* (Mauritania Poverty Reduction Strategy)

 F ish are critically important to the developing world as a source of food for millions and the large economic benefits from employment, export income, and government revenue. Consumption of fish is by far the primary source of animal protein for many coastal, rural communities, where many people in the developing world live.

Fishing takes place on the high seas, in coastal or inland waters on a commercial or artisanal scale.[2] Here we concentrate on commercial fisheries operating in developing countries' waters, which raise important issues at the interface of poverty reduction, resource management and economic rent capture, thereby providing important opportunities for EFR. A particular focus is on "access agreements" between fish-rich, low-income countries and foreign distant water fleets. These agreements mostly apply to coastal nations in west and southern Africa, and island nations in the Pacific and Indian Ocean who could generate significant revenue from these resources. In these countries, moreover, many communities rely on fisheries for their livelihood. Falling stock, and competition between commercial-scale operators and subsistence fishers over the same stocks can undermine the livelihoods of fishing communities.

Key features of the sector: overview

As in other extracting sectors, such as forestry, fisheries resource degradation and depletion are key concerns. Even if complete and irreversible stock depletion remains rare, it may be a concern for the fisheries sector.[3] For now, however, the main challenge in the fisheries sector relates to fish stocks which are exploited in a sub-optimal way, i.e. which could yield more fish with better management. This occurs when excessive fishing constrains stock re-building, leading to a lower than possible equilibrium stock level.

About 75% of the world fishing stock is "fully fished" (47%), "over-fished" (18%), "depleted" (10%) or recovering slowly from depletion (FAO, 2002). A stock is described as "fully-fished" if it is exploited at its maximum long-term sustainable yield. A stock is described as "over-fished" not necessarily because of a threat to its survival, but because it could yield more fish later if the current catches (which include immature fish which could still grow if given a chance) were reduced. This implies that more fish (in terms of total weight) could be caught sustainably by reducing fishing "effort". From an economic point of view, reducing fishing effort in an "over-fished" stock increases the total production while reducing the total costs of production, leading to a much more favourable outcome.

Reducing fishing effort can be done by reducing the number of fishing boats, reducing the capacity of each boat or a combination of the two. How to achieve this reduction depends on the relative importance of various policy objectives. If the objective is to maximise the overall economic value of the catch, it may entail restricting access to a small number of highly cost-efficient fishing boats, whose profit could be shared with the state through taxes, auctioning of access rights or other mechanisms. If, however, employment and the protection of subsistence fishing communities are the primary concern, a larger number of less cost-efficient vessels may be allowed access. If the objective is to maximise food production and (therefore total catch) irrespective of costs, a mid-way solution will be needed.

Pressures on high seas fisheries have accordingly increased in recent years – both from OECD and non-OECD countries. For example, fishing pressure in west Africa has increased significantly since the 1960s from EU, Russian and Asian fleets. Key factors behind this trend include radical advances in fish harvesting technology; ever larger fishing fleets operating in ever-distant waters; direct and indirect government subsidies in many OECD countries, which encourage over-investment in fishing capacity as well as illegal, unreported and unregulated fishing.

Policies and instruments

In the absence of taxation, the financial benefit from exploiting fisheries resources are fully captured by the private sector, without compensation to society at large. In addition individual operators have no direct incentive to restrict their catch, since they do not, individually, derive any direct benefits from doing so. Imposition of levies on volume caught, in combination with proper management measures – which may include restricting access to fishing grounds – can generate revenues to compensate the owners of the resource, (i.e. the country whose fishing stocks are being exploited) and help reduce fishing efforts.

The United Nations Convention on the Law of the Sea (UNCLOS), which sets out coastal states' rights and duties with their "exclusive economic zone", includes provisions for states unable to fully harvest their fisheries resources to provide access to the available surplus to other states, through appropriate agreements.

In many developing countries, "access agreements" are the main mechanism to manage access to national stock by foreign fleets (known as distant water fleets, DWFs). These are negotiated between governments or between the host government and the foreign fishing operator(s). The main DWFs nations are Spain (within the EU[4]), Japan, the United States, Russia, China, Korea and Chinese Taipei. The EU, for example has fishery agreements with 20 developing countries – over half in west Africa – under the Common Fisheries Policy (CFP), aiming to secure access to their stocks and waters for fleets of its member States.

Access agreements generally provide for financial compensation paid by the DWF country (or private operator) to the country in whose waters the fishing takes place. This serves the dual purpose of allocating to the coastal state a share of the profits generated from resource extraction *and* of regulating the harvest. Access agreements sometimes include other provisions such as preferential access to DWF country markets; joint-venture agreements; requirements to include a certain proportion of domestic workers in the crews of the DWF vessels; etc.

Countries which have entered into "access agreements" include some of the poorest and least developed, such as Angola, Guinea Bissau, Mauritania, Mozambique, Sao Tome and Senegal. For some, these agreements represent significant financial resources. For example, it is estimated that EU agreements provide as much as 30% of government revenues in Guinea Bissau, 15% in Mauritania and 13% in Sao Tome (IFREMER, 1999).

While in the past these agreements managed to capture only part of the revenue generated (van Santen, 2001), the situation has been improving over time. One reason is that their provisions have been subjected to increased scrutiny by NGOs as well as within the European Commission. Concerns include the management and exploitation of fisheries resources in developing countries and their impact on the livelihoods of local communities, and the fact that resources from extremely poor countries are, in effect, exploited by foreigners with insufficient compensation. Developing countries themselves have also become more assertive in expressing their concerns over preserving their fish stocks and developing their national fisheries sectors (OECD/DAC, 2002b).

Under pressure from the developing countries concerned and NGOs, the EU fisheries council shifted in 2002 from relatively narrow fisheries agreements to much broader "partnership agreements" which include a range of measures aimed at addressing social and environmental concerns related to the fishery sector in the partner country (see Box 6.2).

Negotiating and enforcing access agreements poses unique challenges. Most importantly, fish stocks may be very mobile and difficult to monitor by developing countries. Countries with limited financial and technical capacity can therefore find it very be difficult, in the context of access agreements, to determine the limits which should be placed on catch in order to ensure optimal management and sustainability. Second, for technological, administrative and other reasons it is difficult, notably for very poor countries, to effectively monitor access by foreign boats and enforce restrictions to fishing. The level of financial compensation secured by the host country depends on its negotiating skills and influence relative to the DWF. Other important factors relate to the capacity of governments to manage complex processes such as the auctioning of access rights in a transparent and open manner and with due safeguards to prevent corruption and other forms of abuse.

Perspectives and interests of main stakeholders

Interests of subsistence fishers communities. Subsistence fishers are amongst "the poorest of the poor". In general, there is a direct link between the volumes caught offshore by commercial vessels and the possibilities for exploiting the same species in the coastal zone. A major concern in such cases is competition between coastal artisanal fisherfolk and domestic and foreign commercial fleets over the same fish stocks. Artisanal fishers are politically marginalized, and typically have little influence on the negotiation of access agreements, so they get only limited benefits.

Interests of commercial domestic fishers. Although subsistence and commercial fishermen often compete directly with each other, both have a common interest in limiting the fishing opportunities provided to the distant water fishing fleets (DWFs) through Access Agreements, *e.g.* by restricting access to particular species, use of certain methods and zones. Namibia is the extreme example of a commercial domestic industry that was built up by strong policies to limit access by DWFs (see Box 6.1). On the other hand, in some countries, which have no fishing tradition, such as Mauritania offering access to foreigners may be the best way to maximise the economic benefits from their rich fisheries resources.

Interests of distant water fishing fleets. Stock fluctuations, over fishing, depleted resources in developed as well as the policies of many DWF countries to subsidise investment

> **Box 6.1. Improved fishery management and increased rent capture –
> the case of Namibia**
>
> Prior to independence in 1990, access to Namibia's fisheries resources was largely uncontrolled and coastal waters were massively over-fished, primarily by foreign fleets. The newly elected government instituted a new policy, legal and management framework to effectively manage its fisheries and develop a domestic industry. Quota fees – based on total allowable catch for major species – and license fees were introduced with fishing rights biased to Namibian vessels. By-catch fees and a Marine Resources Fund levy were imposed, based on tonnage of landed catch to finance fisheries research and training. As a result, the sector contributed about USD 220 million to GDP in 2000 and was valued at USD 354 million in 2001. The indirect benefits have also been substantial: the fish processing industry has grown rapidly. The number of whitefish-processing plants has grown from zero in 1991 to more than 20 in 2002, and employment in the sector has grown to about 14 000 people. The government also invested heavily in monitoring activities, with an integrated program of inspection and patrols at sea (*i.e.* on board observers), on land (monitoring of port landings) and in the air (via satellite). While expensive, this investment pays off. The ratio of monitoring costs to value of landed catch has declined from an annual average of 6% over 1994-1997 to under 4% in 1999, reflecting an increasing value of landed catch. Namibia's rights-based fisheries management system incorporates an effective monitoring and compliance system at a cost that is commensurate with the socio-economic value of the sector. As a result, Namibia enjoys very high levels of compliance by its fishing industry, a situation very different from 1990. In its efforts to improve fisheries management, Namibia has benefited from assistance from several donors.
>
> *Source:* Nichols (2003).

in fishing vessels – leading to overcapacity – combine to encourage a search for new stocks to fish. Moreover, as will all fishing operators, DWF have inherent incentives to fish beyond catch limits and to underreport their catch. In addition the incentives for DWFs to sustain fish stocks in a given area are low as they can move to other countries once stocks are depleted. Therefore, DWF fleets have a clear interest in lobbying for increased access, notably when part of the price of access (*i.e.* various forms of compensations to the host country) is borne by public authorities (*e.g.* the EC).

Interests of DWF countries. For DWF countries, fisheries agreements provide a way to keep their fishing industry in business, to ensure security of supply for downstream industries and to generate or maintain jobs. For the EU, agreements account for 20% of fisheries production, providing both direct employment and many more "ancillary jobs". Spain, for example, which has one of the largest fleets within the EU, is highly dependent on EC international fisheries agreements, since it obtains nearly half its catch in waters outside those of EU member States.

Interests of developing country governments. Coastal States governments have to balance a range of competing objectives in relation to fisheries management: These include maximising food production to enhance food security; promoting exports to maximise foreign exchange earnings, preserving the livelihoods and employment of subsistence fishing communities, promoting the interest of the domestic fishing industry and ensuring the sustainability of the resource. **Fisheries departments,** which generally operate under the ministry of agriculture, are often particularly poorly funded and staffed because of the low political priority given to fishery policy. They generally face considerable resistance from affected stakeholders in their efforts to improve fisheries management, notably when this implies reducing fishing effort and strengthening monitoring and enforcement. Their authority to monitor and control the industry gives rise to rent-seeking opportunities, and so corruption is a concern. **Finance ministries** play an important role in negotiating fisheries agreements, particularly in the final stages. They tend to focus on the financial benefits to be gained from fisheries and the cost-effectiveness of monitoring and enforcement, rather than on the complexities of sustainable fisheries management, which can lead to tensions with fisheries departments.

Managing the reform process: key steps

Strengthening states' bargaining power. Negotiation skills are essential for getting good terms in bilateral agreements. Regional co-operative mechanisms can help. By getting together, states which share fisheries resources can improve their bargaining power. For example, the islands of the South Pacific, which formed in 1987 a joint Fisheries Forum Agency, have been able to negotiate improved terms with DWFs.

Allocating adequate resources for fisheries management. This includes assessing the state of the resource in order to determine optimal levels of fishing efforts, as well as monitoring, control and surveillance of fishing operators. These activities require sophisticated equipment and specialised skills, and are therefore expensive. In the EU, for example, the overall cost of monitoring fishing activities is estimated at around 5% of the value of production. Earmarking sufficient funds to finance these activities is essential. The recently completed agreement between Mauritania and the European Union earmarks funds for enhanced control and surveillance of fishery activities.

An important issue in this regard concerns cost effectiveness of these activities. In India, for example, the Ocean Tuna Commission (IOTC) is developing its inspection and control scheme by systematically analysing all known control techniques and selecting the most cost-effective.[5] Part of the monitoring and enforcement costs are shouldered by the industry either directly (*e.g.* through compliance with documentation and reporting obligations) and indirectly (*e.g.* through licensing fees).

Encouraging DFW nations to improve the coherence of their policies. The fisheries policies of some countries contradict the objectives of their development co-operation policies. Pointing out these contradictions can foster reform. For example, the DAC Peer Review of Spain (OECD, 2002) encourages Spain "... *to consider how to prevent domestic interests from taking precedence over development co-operation objectives when debating the Common Fisheries Policy as well as fisheries agreements in the European Council*". Similar arguments apply to many other DAC member countries with DWF fleets.

Box 6.2. **Hard bargaining for fishing agreements: experience in west Africa**

West African countries have recently stepped up their bargaining efforts with EU and, as a result, secured much more favourable terms. In 2001, Senegal, Mauritania and Guinea Bissau banned EU fishing, and sought to increase regional co-operation on fishery management. Subsequently, Mauritania agreed with the EU a new five-years protocol for fishing – with financial compensation of € 430 million, as compared with € 267 million under the previous agreement. This will raise fishery revenues to about 8% of GDP and close to 30% of total revenues – with some earmarked for developing the local fishery sector, enhanced control, and surveillance of fishery activities (Islamic Republic of Mauritania, 2002).

In 2002, Senegal signed a new agreement with the EU. This decreased fishing possibilities on sensitive stocks for EU vessels and provides a 2-month biological rest to give more protection to fish stocks and to minimise the risk of competition with the artisanal fleet. The financial compensation was also been increased from € 12 million to 16 million per year with a share of 18%, that is € 3 million, earmarked for a range of measures designed to support conservation of fish stocks and strengthen the Senegalese fishing sector. These include support for scientific research and evaluation of the state of fish stocks, development of the Senegalese control and monitoring system, the development of the Senegalese institutional structures for fisheries management, and provision of training for fishermen.

Source: Islamic Republic of Mauritania, 2002; Europa: Gateway to the European Union Press Release 26/06/02: Commission welcomes renewal of EU/Senegal fisheries protocol. *http://europa.eu.int.*

Notes

1. This chapter has benefited from comments by Carl-Christian Schmidt and Bertrand Le Gallic of the OECD Directorate for Food, Agriculture and Fisheries, as well as from Paul Steele of DFID.

2. The FAO (*www.fao.org/fi/glossary/default.asp*) defines artisanal fisheries as "Traditional fisheries involving fishing households (as opposed to commercial companies), using relatively small amount of capital and energy, relatively small fishing vessels (if any), making short fishing trips, close to shore, mainly for local consumption."

3. A well-documented example concerns Canada's "Atlantic Cod" stock, which has not recovered from excessive fishing despite drastic fishing restrictions for several years. There are also countries where some coastal fisheries have been exhausted. However, these remain exceptions.

4. Within the EU the biggest fishing fleets are from, in descending order, Spain, Italy, Portugal, and France. [EC, (2002)].

5. Beslier S. (2004), *Enforcement and surveillance: What are our technical capacities and how much are willing to pay?* In: *Fish Piracy*, OECD (2004).

ISBN 92-64-00868-3
Environmental Fiscal Reform for Poverty Reduction
© OECD 2005

PART II

Chapter 7

EFR in the Fossil Fuel Sector

The Law on the special social protection of certain categories of population that eliminated non-targeted energy subsidies, communal services and limited energy privileges to eleven categories representing the most vulnerable segments of society, is likely to have a beneficial effect on government finances and improve significantly the efficiency and targeting of the government's social assistance program. (Moldova Interim Poverty Reduction Strategy)

Fossil fuels are key inputs in most economic activities. In their various forms, they are used for transport (gasoline, diesel), cooking and lighting (kerosene); heating (heavy and light fuel), pumping water (diesel), mechanical power (diesel) and many other uses, especially where electricity is not available. Fossil fuels are currently dominating global energy use and will continue to do so in the foreseeable future, accounting for some 85% of the increase in world primary demand. Two-thirds of the expected future increase in energy demand will come from developing countries. By 2030, they will account for almost half of total demand, the major part of which will have to be met by fossil fuels.[1] Fossil fuel consumption, in the home and outdoors, is also a major cause of air pollution. Pollution "hot-spots" – such as the cities of Bombay, Jakarta, Manila or Mexico City – are growing. Vehicles on developing country roads, including diesel and gasoline cars and trucks and scooters powered by two-stroke engines, generate considerable harmful pollution. Fossil fuels are also heavy and growing contributors to greenhouse gas emissions. This chapter reviews the scope for EFR in relation to the consumption of petroleum-based fuels, which are a major source of energy for transport, lighting, cooking, heating and industry in all developing countries. (Issues relating to coal or coal conversion into fuel – as done on a large-scale in South Africa – are not covered. Nor are issues relating to oil extraction and refining.)

Key features of the sector: overview

OECD countries tend to tax petroleum-based fuels heavily.[2] The transport sector accounted for 90% of total environmentally related tax revenues, raised from taxes on petrol, diesel fuel and motor vehicles.[3] In some OECD countries, excise taxes on fossil fuel have become the third-largest source of tax-revenues. Motor fuel taxes have stimulated rapid progress in the energy efficiency of motor vehicles.

In developing countries, by contrast, fossils are often lightly taxed or even subsidised. According to a 2002 report by the IMF, UNEP and World Bank: "For the developing world as a whole, the net effect on the public budget from phasing out subsidies to gasoline and diesel could reach USD 18 billion. Moreover, if countries with low taxes on those fuels were to increase them to the average level in their respective region, the net effect would add to some USD 71 billion." (IMF et al., 2002: "Financing for Sustainable Development".) In oil-producing countries, such as Iran (see Box 7.1), Yemen, Venezuela, Nigeria and Indonesia, fuels are often sold at below the cost of production. But even oil importers like Bangladesh, China and India also subsidise fuels, although the latter has in recent years reversed this policy.

In many countries, ending the underpricing of fuel could free up considerable fiscal resources while also reducing waste of energy and helping reduce pollution. This creates a potential for very significant fiscal, poverty-reduction and environmental benefits.

Box 7.1. **Fuel subsidies in Indonesia and in the Islamic Republic of Iran**

In **Indonesia**, the government directly subsidises oil prices which are among the lowest in South-East Asia. These subsidies, which currently absorb more than 10% of the state budget, incur large economic, environmental and social costs. A recent government review of the subsidy policy concludes that eliminating subsidies would reduce government expenditure, increase foreign exchange earnings and reduce environmental damage, particularly from airborne emissions of particulate matter and lead. The cost of the subsidies applied to kerosene, automotive diesel, industrial diesel, motor gasoline and heavy fuel oil amounted to almost USD 4 billion in 2002. It is projected that between 2000 and 2005, a total of USD 36 billion would be spent on oil subsidies if they were left unchanged. In addition, the value of lost foreign exchange earnings caused by lower exports would reach USD 16 billion. Subsidy reform would allow financial resources to be redirected towards supporting the poor in more effective ways, such as through a voucher scheme.

In **Iran**, economic incentives are so heavily distorted, that reforms would yield welfare gains estimated at about 19% of the GDP. A large part of this stems from petroleum prices that are only about 10% of world prices, with an implicit subsidy to petroleum products that amounts to more than 18% of GDP. Such subsidies encourage excessive and wasteful energy consumption, with substantial foreign exchange earnings forgone. Poor people benefit less from these subsidies. Their removal would release vast resources that could be redirected toward environmental, social, and other expenditures, underpinning poverty reduction and sustainable development. The prospects for reform hinge on the opening of the political process, allowing greater voice and participation. The first step in addressing fuel subsidies is to display them explicitly in the budget. This would highlight their magnitude in relation to other priority areas, and facilitate a process of gradually lowering the subsidy to allow fuel prices in Iran to rise to world levels.

Source: UNEP 2004: Energy subsidies – lessons learned in assessing their impacts and designing policy reforms, *World Development Report 2003*.

Policies and instruments

Availing of the "win-win" opportunities mentioned above generally involves facing a number of tradeoffs and applying a mix of regulatory and fiscal instruments.

a) Differential pricing: social vs. environmental concerns

Environmental consideration would call for taxing more heavily the most polluting fuels such as diesel. *Social concerns* call for taxing less fuels used by the poor such as kerosene used for cooking relative to those used by the relatively richer (gasoline for cars). In many countries, social and equity consideration prevail and diesel, which is used for commercial, industrial or farming activities, is often taxed at a lower rate than gasoline which is primarily used by private cars and motorbikes.

The scope for substitution across different fuels is another critical issue to consider. For example, if gasoline becomes more expensive, many scooter drivers will mix it with cheaper kerosene, which sharply increases harmful pollution. Similarly, if kerosene becomes too expensive, and wood readily accessible from forests, households can switch to fuelwood. Besides fostering deforestation, this increases women's burden of collecting fuel and increases exposure to harmful pollutants from burning wood indoors. Where these risks are high, there may be a case for continued subsidisation of fuels used particularly by the poor, until the poor can gain access to modern fuels such as liquefied petroleum gas or electricity which are preferable in all respects (health, safety, efficiency and environment). (See Box 7.2 on LPG subsidies in Senegal.)

Figure 7.1. **Taxation of oil products as % of retail price (example: Kenya)**

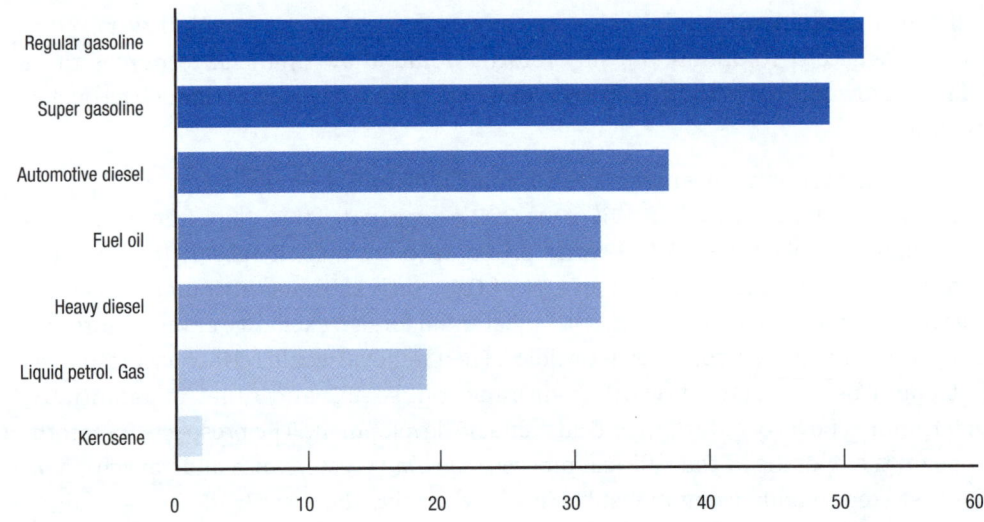

Source: Metschies (2001).

b) Targeted subsidies

The stated reasons for subsidies or low taxes on certain fuels include distributional concerns. Thus, fuels used more by the poorest are kept cheaper. Raising their taxes is often politically difficult. However, in practice these subsidies end up benefiting disproportionately those consuming the largest volumes who are seldom among the poorest.

If increased fuel prices are expected to hurt the poor, subsidies to those likely to be impacted is a much better way to address the issue. For example, the rural poor living in remote areas will benefit much more from policies supporting the expansion of distribution networks for modern fuel or electricity by private sector suppliers, which may involve subsidised credit, than from subsidies on fuels themselves, which are not available to them. Subsidy measures can be designed to be technology-neutral or to support renewable energy and energy-efficient end-use equipment such as stoves, lamps etc.

c) Regulations on the quality of fuels and vehicle maintenance

Fuel taxes alone cannot fully address air pollution issues. Apart from CO_2, most polluting emissions are indirectly linked to the quantity of fuel consumed. Regulations specifying the chemical composition of fuels, (*e.g.* the substitution of lead by less harmful substitute, or maximum concentrations of sulphur) are needed. These are implemented at the level of the refinery. Regulations on the chemical contents of fuels and differential taxation prices must be designed to reinforce each other to discourage most polluting fuels. Regulations on vehicle maintenance standards can also be useful, in countries where administrative capacity is sufficient to ensure enforcement.

d) Integrated urban planning and transport management

The instruments outlined above are strongly complementary with broader urban planning measures which aim to address such issues as congestion, vehicle-related air pollution and traffic accidents and, more generally, improvements in health and quality of life in cities and containment of urban sprawl. These measures include integrated land use planning, the promotion of electrified mass rapid transit systems; traffic management, parking regulations and many others.[4]

Administrative feasibility issues. Fuel taxes are attractive from an administrative point of view as they are relatively simple to manage. They can be best levied at the level of the producer or importer/distribution centre, who are usually few in numbers and easily identifiable geographically.[5] These actors will then pass the tax on to the final consumer. In some regions, however, cross-border smuggling is an issue. Where borders are difficult to control, regional co-operation and harmonization of fuel pricing policies may be a precondition to the successful application of fuel taxation measures.

Box 7.2. Impacts on the poor of LPG subsidies

In Senegal, after subsidising petroleum products throughout the year 2000, the government reinstated the full pass-through for petroleum products at end-June 2001. However, in order to protect consumers when the world price of oil increases again, the authorities have decided to maintain the price of butane gas below the world market price, since the poor predominantly use it. The government is widely distributing butane gas, even to rural areas, with the objective of reducing the use of charcoal and firewood as fuel and, therefore, decreasing deforestation. The subsidy is partially covered by resources under the interim relief granted by Senegal's creditors under the Initiative for Heavily Indebted Poor Countries (HIPC Initiative).

In India LPG subsidies have led to large distortions in the energy markets and rationing. Not surprisingly the subsidies tend to reach the non-poor so that in Hyderabad LPG accounts for 40% of energy use by the 10% most well off households, and only 4% for energy of the poorest households.

Source: Kende-Robb, C. and K. Yao (2003); and UNEP, IEA, 2002.

Interests and perspectives of key stakeholders

The poor. Poorer households tend to spend a higher share of their income on energy than the average. While low fuel prices benefit primarily the relatively well-off, the urban poor also benefit as they rely heavily on public transport (which uses diesel) and kerosene (for cooking and lighting). Poor rural households, who have much less access to modern fuels and rely heavily on biomass, benefit much less. In general, targeted subsidies to the poor are more appropriate to compensate the poor from price rises.

The bureaucracy. Finance ministries favour fuel taxes (or subsidy reductions) because of their high revenue potential and administrative simplicity. In countries struggling with high inflation, however, it is often feared that rising fuel prices might intensify inflation. Other ministries, like those responsible for economic development, energy, industry and transport, are often linked with the interests of related industries and tend to oppose higher fuel taxes. Environmental ministries seldom play an active role in these debates.

Politicians. Elected officials face the pressure from their voters. In many countries, popular discontent (especially from the urban middle classes) has forced the revisions or cancellation of planned tax increases or subsidy removals. This is particularly acute in oil-producing countries, where a history of subsidisations has created entrenched patterns of dependence, and cheap fuel is almost considered a right. In oil importing countries, consumers can more easily understand that oil is a scarce commodity and that wasting it has significant opportunity costs.

Fuel producer and distributors. Due to their economic and political power, in an industry characterised by a high degree of concentration, oil companies are a very influential lobbying group in many countries. They usually oppose higher fuel taxes but are sometimes accused of taking advantage of higher fuel taxes by hiding price increases behind the taxes.

Energy intensive industries. Taxation of fossil fuels raises the costs for key production factors of domestic power plants and energy intensive firms. This may result in heavy competitive disadvantages for the sector if i) substitution possibilities for inputs are limited, ii) the produced goods (like energy, manufactured steel, etc.) are traded widely in the international market and/or iii) trade policy measures like border tax adjustments cannot be applied due to international rules. These industries often seek, successfully, exemptions form price increases. Agriculture is also typically exempted from fuel taxes. In these respects, the situation is similar in OECD and developing countries.

Other industries. The impacts of higher fuel prices on businesses depends highly on the sectoral structure and consumption patterns of the economy. In countries with a history of cheap fuel, there will be strong resistance to price rises. Employees of these businesses share the same interests – jobs and wages depend on the companies' profits. Certain industries, such as public transport sector and rail may benefit competitively from high fuel taxes. Firms specialising in energy efficiency or renewable energy will also find new opportunities.

Designing and managing the reform

As fuel price increases will be negatively perceived by most stakeholders, accompanying measures are critical to ease political resistance:

Acceptance building. Raising awareness of the negative impacts of cheap fuel prices is essential. In oil importing countries the increased dependency, national and direct foreign exchange costs resulting from wasteful consumption should be emphasised. In oil exporting countries, the prospects of switching to oil-importing status within consumers' own lifetime

ENVIRONMENTAL FISCAL REFORM FOR POVERTY REDUCTION – ISBN 92-64-00868-3 – © OECD 2005

can be a powerful argument. Increased awareness of the health impacts of air pollution and information about energy efficiency options can also be convincing, notably to middle class consumers.

Publicising the size of financial savings and their intended uses. This is essential to generate public support and to raise awareness among the "winners" from the reforms. Partially earmarking revenues for social and environmental purposes, notably if they contribute to mitigating the negative impacts of the reform, can greatly promote acceptance of subsidy reductions or tax increases. This may include, for example, earmarking some funds for public transport, urban planning, renewable energy and the like. Revenue may also be allocated to purposes unrelated to transport or environment, *e.g.* to finance general poverty reduction related programmes.

Box 7.3. **The difficulties provoked by fuel price hikes in Indonesia and Thailand**

Indonesia was among the countries hit hardest by the Asian financial and economic crisis. The conditions within the subsequent IMF adjustment programme included the phase-out of fuel subsidies in Indonesia by 1999. When in May 1998 prices for petroleum products in Indonesia rose by over two-thirds, widespread and violent protests forced the government to reverse the price increases. There is no doubt that energy subsidies are a major drain on the government budget. The issue, however, is politically highly sensitive. Due to the Asian crisis large parts of the population found themselves suddenly impoverished and have to struggle for survival. But Indonesia, which is expected to become an oil importer around 2010, will have to face this issue sooner or later.

Source: IEA, 1999.

Coalition-building to phase out unleaded petrol through price reform in Thailand

In 1991 the Government of Thailand – pressed by concerns about the seriously harmful effects of lead pollution on the population and the environment – embarked on an ambitious program to phase out the use of leaded gasoline. This was a complex task, impacting on many sectors. However, the Thai policymakers managed to surmount the obstacles encountered and successfully completed the process in four and a half years, one year ahead of schedule. A crucial success factor was reliance on fiscal incentives to favour unleaded gasoline. To encourage the switch to unleaded, the retail (pump) price was set at B 0.3 (USD 0.012) per litre less than that of leaded gasoline.

This policy was introduced with a collaborative approach involving key stakeholders, such as government agencies, representatives of oil companies, and automobile manufacturers. Success was crucially dependent also on governmental institutions taking vigorous leadership and managing all steps of the process, including setting target dates for implementing key actions, and continual monitoring and follow-up evaluation.

Source: Sayeg (1998).

Targeting compensatory measures. Earmarking revenue for subsidies targeting poor households can both contribute to mitigating the impact of reforms and help build support. This may include support for the expanded access to modern fuel and electricity in rural areas.

Seizing on opportunities. In oil-importing countries, a foreign exchange crisis can increase understanding of the true costs of cheap fuel, and increase acceptability of reform. Conversely, declining world oil prices can create windows of opportunity for introducing taxes or reducing subsidies while, keeping consumer prices constant.

Pre-announcing and phasing-in reforms and sequenced reforms. Sudden and unexpected price hikes generate the most discontent and economic disruptions. The more time consumers have to adapt to higher fuel prices, the lower the perceived and real adjustment costs. Announcing reform steps well in advance even stretches the time horizon and prepares consumers. Step-wise increases in prices are more easily accepted.

Notes

1. International Energy Agency (2004): World Energy Outlook 2004.

2. Coal is an exception to this general rule.

3. An overview of Green Tax Reform and Environmentally Related Taxes in OECD Countries, OECD (2001).

4. These issues are discussed in detail in OECD (2000): Shaping the Urban Environment in the 21st Century: A DAC Reference Manual on Urban Environmental Policy. Available from *http://www1.oecd.org/dac/urbenv/*.

5. Even though fuel taxes are not import duties, they are usually collected by the customs authority.

ISBN 92-64-00868-3
Environmental Fiscal Reform for Poverty Reduction
© OECD 2005

PART II

Chapter 8

EFR in the Electricity Sector

Until recently the government was using the energy sector as a means for providing social protection to the population… unlimited provision of electrical energy allowances were adopted for (certain) categories. (Georgia Interim Poverty Reduction Strategy)

Electricity is used for lighting, communication, motive power (including for small-scale labour-intensive cottage industries) and for agriculture (pumping water). In urban areas, it is also used for lighting, cooking and refrigeration, and powers modern mass rapid transit systems. Today 1.6 billion people or 27% of the world population still do not have access to electricity – 99% of these people live in developing countries, 80% of them in rural areas. In many African countries, not even 10% of the population have access to electricity (UNDP, 2002). Access to electricity is a key factor of welfare and a precondition for industrialisation and economic growth. This is particularly true for women and girls, who have responsibility for many labour-intensive tasks (*e.g.* grinding and processing food) which can be electrified. Improved indoor lighting is also important for education. Development strategies for the sector usually focus on improving the efficiency of existing systems and expanding supply to more households. EFR is a key element of both.

Key features of the sector: overview

In many countries, electricity generation and transmission is controlled by public monopolies. This has often been associated with poor management practices and insufficient or badly targeted public investment. In addition, tariff levels considerably below cost-recovery levels – as well as *de facto* free supply to government agencies – have promoted wasteful energy use and restricted access to electricity services to the relatively well-off urban consumers. As most electricity is produced through the combustion of fossil fuels, notably coal, another effect has been to generate high level of harmful air pollution. In 1999, according to IMF/UNEP/World Bank estimates, "*... the developing world subsidised electricity at a rate of 4% for a total subsidy of USD 102 billion, or 2% of the developing world's GDP*". A large share of this amount is attributable to countries of the former Soviet Union, where access to electricity is widespread and subsidies amount to almost 14% of GDP. In sub-Saharan Africa, where access to electricity is low, and Latin America and the Caribbean, where the sector has been reformed, subsidies account for around 9% of total subsidies.

Below-cost provision of electricity prevents producers to modernise or even maintain production and transmission facilities, let alone expand the power grid to serve more households. Poorly managed or maintained power generation or transmission equipment also leads to a high degree of "technical loss", *i.e.* pure waste. The quality of service is also poor. Power outages, scheduled or accidental, are frequent and low-consumption households are usually the first to be disconnected. Even in urban areas, people may have to fall back to traditional energy sources such as kerosene, charcoal or wood to meet their energy needs.

Policies and instruments

Opening the industry to private investors. The current trend in many developing countries is for the government to open the power generation and transmission to private sector investment. Reform have often been triggered by macroeconomic crises (*e.g.* in Argentina, Indonesia or Bulgaria) and/or pressure from external donors (*e.g.* India, Indonesia, Bulgaria and Ghana). They require formulating a conducive fiscal and regulatory environment for private sector investment in the energy sector (including independent energy production and distribution companies) in order to foster expansion of the network to communities which do not have access. The challenges involved are beyond the scope of this paper, but, in a nutshell, it implies for the government to switch from a role of direct or indirect provider of electricity to one of regulating private industry and enforcing regulations.

Full cost electricity pricing. The bulk of electricity subsidies take the form of provision of energy below cost-recovery levels. The resulting loss of revenue through energy subsidies is a constant drain on the government budget. Although targeted subsidies to low-income households may have merit in some cases, broad-based subsidies usually have negative overall fiscal and environmental impacts and fail to efficiently achieve poverty reduction goals. In some countries, there may be a risk that poor households may switch back to biomass in response to increases in electricity prices. While this is unusual, it has to be considered insofar as the results are negative in terms of health and safety – notably for women and children – and from an environmental perspective.

Improved tariff collection. Reducing under-reporting of consumption, illegal connections, not paying electricity bills, and other forms of theft are important ways to increase revenues of power supply and distribution companies. Toleration of theft works like an arbitrarily targeted subsidy, benefiting law-breakers among rich and the poor alike, and among governmental agencies.

Targeted explicit and transparent subsidies. This implies converting "invisible" off-budget subsidies into on-budget subsidies, *e.g.* lump-sum transfers. These should also be targeted to those who really need support. In general, the case for subsidising the initial costs of connecting to the network is stronger than the case for subsidising actual electricity consumption. However, for both new and existing customers, there may be a case for subsidising service through special, low tariffs – often called lifeline rates – for poor households, defined by income or consumption. Lifeline rates, if used, should be limited to modest levels of consumption such that subsidies are targeted effectively and not appropriated by richer households.

Support for rural electrification. Improving access to the services by the rural poor is a more effective way to help them than subsidising a service which they cannot obtain. In addition, evidence suggest that even the poor can actually afford to pay for electricity and are willing to pay for relatively expensive options such as off-grid renewable electricity, which are often the only available solution. This is both because of the very high marginal value of the first few units of electricity – for such basic uses such as lighting and pumping – and because their total consumption of electricity is modest. Options to support investment in rural electrification may include subsidies on the one time capital costs of expanding service provision, time-bound tax incentives for energy services companies and others.*

---

* The World Bank (2000) Energy Development Report 2000: "Energy Services for the World's Poor" provides a comprehensive analysis of these issues.

It should be noted, however, that improving access to energy in rural areas does not necessarily have to mean electrification, given that electrification is not always feasible or financially viable. In such cases the focus should be on improving access to modern fuels (*e.g.* charcoal briquettes, LPG) and energy-efficient fuel stoves which help avoid waste of energy, lead to time and financial savings for households, reduce accidents and harmful indoor pollution. Women, who suffer the most from lack of access to modern energy, stand to benefit the most.

Reducing high import tariffs on energy equipment. Many countries impose high levels of taxes and import duties on energy equipment, which discourage the adoption of modern, highly energy efficient technologies. Reducing such taxes can help mitigate the impact of increased electricity prices.

Demand-side management measures (DSM). Consumers actually want energy services (*e.g.* lighting, cooking, heating) and not energy itself (*i.e.* kilowatt-hours). DSM refers to administrative, regulatory and technical approaches which allow to provide the same service using less energy. Examples include differentiated tariffs which reduce "peak" demand, helping better manage supply over time. This reduces the need for "peak time" reserve power-generation, cutting down the investment needed to keep the system functioning properly. It also yields significant fuel savings since meeting peak-demand often involves switching on some of the least efficient power stations. The results are cost savings for the provider and the consumer. Promoting the adoption of energy-efficient devices, such as energy-saving lamps, refrigerators and air conditioners by final consumer are other important DSM measures. Efficiency standards imposed on producers of electric appliances and equipment are also used in most OECD countries. All these measures can help compensate the impact of higher prices.

Interests and perspectives of affected stakeholders

The poor. In many developing countries poor households often do not have access to electricity. The rural poor, in particular, are much less affected by electricity price increases than by prices increases of modern fuels. Where significant number of poor households are affected, targeted subsidies, such as "life-line tariffs" (whereby the first few Kilowatts are very cheap or and increase steeply thereafter), can be used. Such approaches are far less costly than untargeted subsidies and can even involve cross-subsidies from heavier consumers. However, they require metering of electricity, which may not be worthwhile nor feasible in the shanty settlements of many developing countries. In most cases, the rural poor will be most interested in any measures that allow them to have access to services.

Non-poor households. Households benefiting from electricity services clearly suffer directly from prices increases. If the quality of those services are highly unreliable, however, and if price increases lead to improvements of initially unreliable electricity, even they may ultimately benefit.

Politicians. By granting subsidies to certain sectors (*e.g.* mining, agriculture), politicians can win key "vote blocs" (*e.g.* miners, farmers). The commercialisation of the electricity industry thus eliminates important discretionary powers and sources of political patronage.

The Administration. The **ministry of finance** is very likely to be an important political driving force supporting subsidy reduction or their conversion to explicit on-budget subsidies. **Parastatal electricity producers and distributors** will often resist any reforms, and notably privatisation.

Industries. Energy-intensive industries enjoying low price electricity may be negatively impacted by reform. However, in many countries, commercial users are often charged significantly higher tariffs to offset below-cost supply to consumers and agriculture. In India for example, the average tariff for industry amounted to 2.9 Rupees/kWh in 1997/98, compared to 1.3 Rupees/kWh for households and 0.3 Rupees/kWh for agriculture, implying that industry partly financed governmental subsidy policies (IEA, 1999). Some enterprises even switched to auto-production of electricity due to these high electricity tariffs. Sectors who bear a high share of the costs of subsidies are natural allies against subsidisation. Equally important for business are reliability and quality of the service. Frequent power outages are very costly requiring investment in expensive back-up systems. If power failures are a problem, energy service improvement through subsidy removal will in the medium term meet widespread approval among the actors of the private sector.

NGOs. Civil society may be ambivalent about price rises. Environment NGOs will support renewable energy production through earmarking of taxes, but may be sceptical to EFR in a context of energy privatisation and favour decentralised renewable off-grid electrification over power-grid expansion. Many environmental NGOs militate for a switch towards renewable energies, even if this requires subsidisation.

Academic Institution and Research Groups. Electricity subsidies are generally implicit, indirect and invisible. Complex analysis is required to estimate accurately their actual magnitude. This is an important role for research groups.

Managing the reform process

a) Publicising the true costs of cheap electricity

As in the case of fuel price reform, it is critical to advertise the actual impacts and costs of low electricity prices in order to overcome political resistance to price changes. People must know the role of electricity subsidies in the government's budget difficulties, notably in times of fiscal crises and understand why power outages are so frequent and service erratic or why they only receive electricity once a week. In some countries, publicising the extent of non-payment by official entities and the fact that low prices benefit the better off can also help.

b) Making full use of available compensatory measures

In general, reforms do not imply eliminating but rather reforming subsidies, notably to ensure that they benefit those who need them. Targeted subsidies, and the other policy tools outlined above can all help compensate for the negative impacts of reforms.

c) Pointing to the beneficial uses of the funds freed up by reform

Recycling revenues for expanding the grid, improving existing facilities or supporting decentralised power supply system for increased access to power for the poor are all measures which become possible with price reforms. Intended uses of the resources freed up by reform should be announced in advance and widely advertised. Estimates of the cost of a well targeted electricity-subsidy programme to facilitate access to the poor in India suggest that this would cost the government around USD 1.1 billion (approximately 47 billion rupees: 11 billion rupees for the connection of new customers and 36 billion rupees for the consumption of poor households). Current subsidies to households and agriculture are much higher, over USD 8.5 billion (365 billion rupees).

Taking advantage of circumstances. Crises and external driving force (*e.g.* an economic crisis, donor pressure, requirements for obtaining private investment from abroad etc.) should also be ceased upon as they provide windows of opportunity for reforms. As in the case of fuel subsidies, reform should be planned and proceed step-by step.

Box 8.1. **EFR in the power sector: experience from Ghana, Argentina and South Africa**

In Ghana, in May 1997, the Ministry of Mines and Energy attempted to raise energy prices by 300% and was met with uproar. The president personally intervened to roll back the increase. Instead parliament was summoned to set up a Public Utilities Regulatory Commission (PURC) in late 1997. The PURC was able to pass in 1998 a similar 300% price increase with much less popular dissent. PURC staff partly attribute this to a concerted public campaign – including workshops, public forums and a media campaign – prior to raising tariffs. The key aim was to persuade consumers that the revenues generated by the price rise would be used responsibly to increase access by the poor.

In the urban areas of Argentina, privatisation and increased user charges initially led to those least able to pay – the illegally connected "colgados" (hangers) in urban slums – being disconnected. Electricity losses of 27% (pre-privatisation) were drastically reduced. This led to public outcry and several court cases were brought on behalf of the "colgados". With mounting media coverage and public pressure, the federal government, provincial government of Buenos Aires and two private distribution companies entered into a "Four year framework agreement". The companies were reimbursed for unpaid balances by illegally connected shanty towns and subsidies were provided for establishing collective meters. In turn, companies agreed to waive any claims on unpaid bills since 1992 and to install 10 000 meters a month in low income areas. As a result of the framework agreement, roughly 650 000 users were formally connected to the network.

In 1991, only 33% of households in South Africa had access to electricity. By 2003, close to 70% of households received electricity. Yearly connection rates have averaged around 450 000 over the last 10 years. Until 2002 the electrification programme was funded by the state utility (Eskom) and local governments. However, since Eskom became a taxable entity, the programme has received "on-budget" funding. This route was chosen in favour of deepening cross-subsidies in the electricity pricing system, and is in line with the National Electricity Regulator's broad policy goal of making electricity prices more reflective of supply costs.

Source : WRI, 2002; IEA 1999, 2002; Datt, Garg and Narang, 2003; and C. Morden (2003), "Environmental fiscal reform: Options for South Africa", presented at OECD Scoping Workshop, Paris, January 2003.

ISBN 92-64-00868-3
Environmental Fiscal Reform for Poverty Reduction
© OECD 2005

PART II

Chapter 9

Drinking Water

Water consumption in the country has dramatically increased due to extremely low prices for water and consequent lack of incentives for consumers to save it, and a lack of effective mechanisms of water resource utilisation and management. (Tajikistan Poverty Reduction Strategy)

Key issues in the sector

Improving access to safe water is amongst the key dimensions of the global fight against poverty and figures prominently among the Millennium Development Goals. This chapter examines the scope for EFR in relation to drinking water, focusing on its political economy aspects.[1] The approaches and instruments discussed in this chapter are relevant and applicable to a wide range of situations and can be considered by countries at various stages of development.

Safe drinking water delivered to households or communities is not a simple good, but the combined result of management of upstream waterways, large-scale investment in pipes, treatment facilities and associated public works, as well as their day-to-day management and maintenance. The cost of water has thus two main components. The *capital costs* of establishing the system (*e.g.* the pipes, water treatment plant etc) and the *running cost* of the system once established (*e.g.* labor costs, costs of chemicals for treatment, etc.).

Financing water provision

Water is a basic good. Access to water can therefore be considered a human right, to be provided free of charge. In most developing countries, water is provided to users at very low cost or even for free. Provision of water at less than cost is generally unfair, insofar as it benefits those already connected to the system who are seldom the poorest. In addition, water is also an input into a range of "for profit" productive activities, and it is not clear that subsidising such activities benefits the public interest. It also encourage waste. Moreover, unless these shortfalls are fully covered though subsidies, below-cost provision prevents water utilities from maintaining the systems, let alone expand them to serve more users. Water supply systems therefore deteriorate over time.

Since providing clean safe water is not free, it has to be financed in one way or another. If subsidies are to be provided, how they are financed should be made explicit. Subsidisation through below cost provision of water is neither transparent not targeted towards the needy.

There is, therefore, a wide scope for win-win reforms to i) reverse the deterioration of infrastructure, ii) mobilise funds to allow expansion of water supply systems to poor communities, iii) reduce strains on the public budget, and iv) reduce waste of water and preserve the resource for the future. Water pricing can also play an important role in reducing the dependence of water supply companies on public authorities while also encouraging water users to play an active role in monitoring the quality of services provision and demand "value for their money", rather than behave as passive recipient of a service.[2] The vital need for improved cost recovery from users is among the key conclusions of the World Panel on Financing Water Infrastructure (the "Camdessus Report").

ENVIRONMENTAL FISCAL REFORM FOR POVERTY REDUCTION – ISBN 92-64-00868-3 – © OECD 2005

Water pricing is, however, politically controversial. As with all pricing reforms, the poor may be impacted negatively although there are however many ways of avoiding this, notably through targeted subsidies. There are also legitimate concerns relating to abuse of monopoly power by water supply companies. The information asymmetry between governments, consumers and water utilities on the other leaves the sector vulnerable to social criticism and political interference. For all these reasons, water pricing must be considered as part of a broader package of institutional reform. Key challenges include ensuring "value for money" when public funding is used to support access to safe water and devising cost recovery systems which avoid negative impacts on poor and vulnerable groups.

Policies and instruments

In many countries, urban water and sewerage systems are managed by municipal or district water companies owned by local authorities. Water is provided at prices well below long-run financial and environmental costs, resulting in overuse and waste. High levels of uncollected accounts and system losses due to poor maintenance and outright theft through illegal connections accentuate this. Combined with poor management practices, these undermine the ability of public water utilities to maintain, let alone expand or upgrade, their network. Water and sewerage services, often confined to relatively high-income groups, are irregular and unreliable.

Many cities urgently need to comprehensively reform policies and institutions to stop the rapid deterioration of water infrastructure, to promote efficient and sustainable use of water, and to generate revenues for needed investments. In addition to increased cost-recovery, this generally requires improved resource conservation and fostering pollution prevention at the source. It also often implies mobilising private capital and management expertise to finance and operate water supply infrastructure. These imply reforms in a number of areas:

Regulatory controls. A key aspect of reform is the separation of the functions of water provision, and that of monitoring the performance of water utilities, which enjoy a *de facto* monopoly over a given service area. Regulatory oversight is critical to ensure that utilities – whether public or private – perform properly. This includes establishment of minimum standards regarding the quality of services; and establishing who gets access to services and under what conditions. A key issue is to ensure that shifts towards full cost recovery do not unduly penalise the poor.

Legal statutes of water utilities. These must provide the basis for their financial autonomy and accountability, notably to allow them to raise additional finance to cover the capital costs of network improvement and expansion. They also must enable them to exert control on upstream activities, such as pollution, which affect the quality of the water resources and therefore the cost of delivering safe water. They must also allow for transfers of management functions to user organisations or the private sector on a commercial basis.

Monitoring of water utilities. This concerns in particular operational and financial performance in areas such as billing and collection efficiency; control of operating costs; transparency in relations with the authorities, clients and media and capacity to plan and carry out complex capital investment projects, for example in the context of public-private sector partnerships.

Realistic financing strategies. National public funding is expected to remain the main source of finance for the water sector for the foreseeable future. Investments should therefore be based on sound financial analysis, taking into account the financial constraints faced by public budgets over the relevant time frame, the willingness and ability to pay of customers, and prospects for securing external finance. The formulation and assessment of investment plans should be transparent and open to public scrutiny, in order to raise awareness of the costs involved and increase acceptance of the need for cost recovery. The FEASIBLE model developed by the OECD and Denmark can be a useful tool in this respect (see Box 9.1).

Water tariffs. Tariff levels must provide sufficient incentives for the efficient use of water and discourage excessive use. They must also enable the development of sustainable financing systems that will enable water services to be provided on a commercially viable basis (taking into account affordability considerations). These include user charges linked to consumption levels and charges on effluent disposals.

Improving collection efficiency. Sizeable cash flows can often be generated from users by simply increasing collection rates, and making billing systems more transparent, reliable and user-friendly.

Box 9.1. **What water systems can countries afford? Matching plans and resource availability**

One critical barrier to developing water supply systems is the mismatch between what governments investment plan and citizens demand, and their actual costs and affordability. The Danish Government and the OECD have developed an approach to address this issue, in the context of eastern Europe, Caucasus and central Asia. This approach, backed by a special decision-support tool called "FEASIBLE", allows an assessment of the costs of different options, based on technical parameters, and their financial feasibility under different scenarios regarding levels of charges, interest rates, public subsidies, etc. This is complemented by participative dialogues engaging the major stakeholders. The analysis generally reveals financial deficits which would likely undermine implementation of plans. The objective is then to find ways to bridge this gap, either by mobilising additional resources, scaling down the plans and/or extending the time frame. This approach is being tested in the Sichuan province of China, in relation to wastewater collection and treatment infrastructure investment plans. "FEASIBLE" analysis revealed significant budget shortfalls and a need to mobilise additional resources. It also recommended increased user charges, as 80% of the population could afford to pay more, as well as direct income subsidies for the poorest 10-20% of the population.

Source: OECD (2003), *Financing Strategies for Water and Environmental Infrastructure.*

Perspectives and interest of key affected stakeholders

The poor. Poor people spend proportionately far more of their income on water than richer groups. There is a widespread perception that increased cost recovery for water will hurt them the most. However, those who do not have access to water supply services often have to rely on informal water vendors who charge prices many times higher than piped water rates. As in the case of energy (covered above), the poor stand to gain from any reform which actually improve their **access** to the service. Thus, the poor stand to gain considerably if improved cost recovery on water provision allows the expansion of water supply and sanitation services.

These perceptions that higher prices hurt the poor can detract attention from another critical issue: unlike electricity, the provision of water and sanitation implies considerable public works, and investment in fixed infrastructure, such as underground pipes. This raises special issues for informal urban settlements (shanty towns) which are often illegally established on private or publicly-owned land. The key barriers to obtaining and paying for water services may be linked to insecurity of tenure to land rather than general willingness or ability to pay for water. Lack of clear property rights, uncertainties as to the ownership of various plots as well as unplanned development makes it extremely difficult, for legal as well as technical reasons, to develop the infrastructure, while also reducing sharply the willingness of households to do their share of the work and investments required. The formalisation of illegal or unauthorised settlements raises very difficult political issues, beyond the reach of this document. However, these issues are at least as important in conditioning access to water and sanitation by the urban poor, as those linked to the price of water.

Interests of non-poor. Non-poor households are often the main benefactors of under-priced water since they have better access to subsidised water services (in some developing countries more than 80-90% of richest fifth of population compared with 30-50% of poorest fifth).[3]

Water vendors have a clear interest in maintenance of status quo, i.e. of poor and unequal access, because their livelihood depends on the arbitrage opportunities this generates.

Private water companies. International water supply companies, while keen to invest in new markets, have come to appreciate the high risks involved. In particular, they have little scope for protecting themselves against the exchange rate risks, which is critical insofar revenues are exclusively in local currency, while the cost side often includes foreign currencies, partially through international loans. Typically they will prefer to operate a publicly-owned network for a fee (concession contracts) rather than seek ownership of the infrastructure. In most cases, they will seek risk-sharing arrangements as a condition for entry. Thus, it is commonplace that government or donors provide initial funds, as well as provide various forms of risk protection. The profit-motive implies that the private sector, whether national or foreign will have a tendency to concentrate on the most profitable parts of a water system to detriment of poor/isolated regions. On the other hand, since private firms tend to be more efficient in providing services and collecting bills, they can in some cases extend services to the poor more effectively than under public ownership.

Interests of politicians. Local level politicians often seek to increase constituency support and build political power by opposing water price rises. A tradition of low water prices, combined with lack of awareness of the actual costs of "cheap water" on the part of consumers, make it very easy to politicise the issue of water prices. At the national level, however, politicians must balance various sectoral interests and may be more concerned about the drain on public resources resulting from water subsidisation, as well as about the low level of coverage of the poor under high subsidisation.

Interests of public administration. As in the case of energy, finance ministries have a clear interest in stemming the drain on public finances resulting from implicit subsidisation of water. For the authorities responsible for water supply (ministries of public works or municipal water authorities), the shift from directly providing water infrastructure and services towards regulating services provided by private or public utilities and monitoring performance is very significant. And, it may be resisted out of fear of loss of control, power and authority, jobs and opportunities for bribery. Health ministries do not generally play prominent roles in debates regarding reform in the water sector, although the impact on public health can be very significant.

Fostering the reform process: some key steps

a) Making the cost of "cheap water" visible

As with other subsidy schemes, the benefits of water subsidies are visible for beneficiary groups while the costs are diffused broadly and easy to hide. The social, financial and environmental costs of cheap water must be better publicised, to foster a willingness to reform. Quantification of the various implicit subsidies and other support measures for the sector, and their impact on national budgets is essential. The benefits of financially and environmentally sound water supply and sanitation management, in terms of improved quality of water and reliability of service and the resulting health benefits must be emphasised. When implicit subsidies are proposed to be replaced by explicit assistance targeted at the poorest, this must also be made clear and transparent.

Publicising the level of service discrepancies among providers in a given city or region, ("benchmarking") can also encourage providers to improve their performance with regard to prices and efficiency, and help individual consumers and associations in their efforts to demand better service.

b) Establishing strong regulatory systems

Regulatory authorities are responsible for ensuring the public accountability of water utilities, and protecting users against abuse of a monopoly position. The establishment of a credible regulatory authority is a precondition for reform.

c) Building in explicit compensatory measures to protect the poor

In many cases, there will be a case for continued subsidised public investment in water supply infrastructure to allow expansion of the network to more users. The issue is not so much whether to subsidise the sector but how much, and for whose benefit. When significant parts of the population has no access to water services, *subsidising investment in infrastructure* needed to expand access, is likely to be much more effective rather than subsidising the water itself. Such subsidies are effectively self-targeting. "Geographic targeting", whereby an entire area with a concentration of poor households benefits from subsidies, can minimise the administrative costs of managing the subsidy while minimising the extent of "capture" of benefits by the non-poor.

However, among those who do have access to water supply services, the poorest may well be impacted by moves towards increased cost recovery. There are many ways to compensate for that. They include cross-subsidies, differentiated ("life-line") tariffs for the first few units consumed and direct income subsidies to the poor consumers (see Box 9.2). These should be *clearly targeted* and provided in a transparent way. When poor households share water connections, however, such tariff structure may effectively push them towards the high consumption-high tariff range.

Box 9.2. Assessing the affordability of water tariffs

Experience in the countries of Eastern Europe, the Caucasus and Central Asian countries, some of which impose relatively high charges on users, shows, firstly, that affordability is a serious problem for a relatively small share of the population (10-30%). And, secondly, that these most affected social groups are often well defined, *e.g.* pensioners. Replacing existing price subsidy schemes with more targeted income support for specific social groups would be more efficient and result in overall savings in public budgets.

---

d) Breaking the vicious cycle of "low quality – low willingness to pay – low revenues – low quality"

There are strong links between quality of the water service and success in collecting water bills. When service is poor, many consumers feel justified to delay or avoid paying their water bills, further undermining the system's ability to improve quality. Conversely, improved service is generally followed by higher rates of collection. Until the quality of service can actually be improved and consumers have noticed it, the transition period can be difficult.

Box 9.3. Breaking the vicious cycle in Guinea/Conakry

In the city of Conakry, in the West African state of Guinea, creative financing was used to help break out of the vicious cycle of "low quality-low willingness-to pay-low revenues". In 1987, the government water utility functioned very poorly, and the quality of services in Conakry was abysmal. The government of Guinea decided that they wanted to attract the private sector, an approach which had worked well in the Ivory Coast and other countries in the region. The problem was an obvious one – no private company would be interested in a contract when revenues were only a fraction of the costs! The solution found has broad applicability: the private operator was assured of sufficient revenues by a combination of (initially low, but rising) revenues from users and (initially high, but declining) subsidies from the government (largely paid out of a World Bank credit). This time-bound, transparent "transition subsidy" helped improve services, paving the way for subsequent tariffs increases, which could finance the improved service. The vicious cycle was replaced by a virtuous cycle of good service and reliable revenues.

Source: J. Briscoe, the World Bank: "The Challenge of Providing Water in Developing Countries".

---

e) Sequencing cost recovery measures

The pace of increase in water tariffs towards cost recovery levels should be gradual. A first target should be to cover operation and maintenance costs, gradually increasing to recover capital investments and ultimately reflecting environmental costs as well.

Notes

1. Broader and equally critical issues relating to the management of water resources, ranging from the control of water pollution in upstream watersheds, exploitation of non-renewable water, allocation between agriculture, industry and urban uses, and cross-border water flows are not considered here.

2. "Making Services Work for Poor People", *World Development Report 2004* explores these issues in depth and they are therefore not covered here.

3. De Moor (1997).

ISBN 92-64-00868-3
Environmental Fiscal Reform for Poverty Reduction
© OECD 2005

PART II

Chapter 10

Fiscal Measures for Industrial Pollution Control

Pollution of water, air, and soil has severe health impacts. The poorest households are most directly exposed: at home through consumption of contaminated water or food such as fish and shellfish; outside through exposure to health-threatening air pollution; and at work through direct exposure to hazardous substances. Water pollution can also increase the cost of operating water supply systems.

The main forms of pollution include the emissions of harmful or toxic gases in the air, the release of effluents in waterways and the uncontrolled generation of solid wastes. They are generated by industry agriculture and households. This chapter concentrate on pollution from formal industrial-scale activities, which can be effectively addressed by EFR instruments. Issues related to pollution from small-scale "informal" industries, which sometimes use highly toxic inputs (for example, metal-plating), generally call for different instruments than EFR and are not covered here.

Key issues in the sector

In the context of rapidly urbanisation and industrialisation – two critical features of development – early measures to prevent pollution and contamination before they become uncontrollable are critical. Countries such as China, which has enjoyed rapid growth, as well as Eastern European countries in transition, have come to appreciate the serious threats and costs associated with uncontrolled pollution, and the huge cost of subsequent remedial actions.

Traditional approaches to pollution prevention focussed on regulation, for example by imposing pre-determined maximum levels of pollution backed by sanctions, such as fines in case of non-compliance. The disadvantage of these approaches is that they do not encourage firms to exceed standards, even if technological progress reduces the cost of doing so. Aside from penalty avoidance, firms have no incentives to comply.

From the 1990s onwards, there has been a growing recognition of the need to complement regulatory with so-called "market-based instruments" to provide financial rewards for pollution prevention and abatement efforts. Regulatory and market-based approaches are not substitutes but mutually-supportive complements. Both require effective monitoring and enforcement capacity, to set appropriate emission standards, identify emissions in excess of those standards and collect the corresponding taxes or fees. They can only work if sanctions against violators (such as failure to pay effluent taxes or charges) are upheld by the judicial system. Increased awareness of the health and other impacts of pollution among the public, through information and disclosure obligations can also complement this, through public and political pressures. A key challenge is to combine all these approaches to achieve maximum effectiveness.

Policies and instruments

Environmental taxes are levied on harmful industrial inputs (such as high-sulphur coal), emissions and wastes, thereby encouraging pollution prevention and reduction up to the point where the associated costs are greater than the tax. Such taxes can raise revenues, including for environmental monitoring and enforcement activities. Experience from China, the Philippines and Columbia suggest that pollution taxes can generate rapid large and sustained decline in industrial emissions (World Bank, 2000).

Environmental charges, also levied on pollutants and wastes, aim to recover the cost of their safe management and disposal, for example, through waste water collection and treatment facilities. In many countries, such critical environmental services are provided below cost recovery and financed by municipalities' general budget. Improved cost recovery through environmental charges can directly reduce the budgetary burden.

Key principles and approaches in applying these instruments include:

Taxing "bads" rather than "goods". It is generally preferable to tax actual pollutants (*e.g.* sulphur emissions) rather than inputs or products which are associated with the generation of pollution (*e.g.* fuel which contains sulphur). Similarly waste collection treatment charges should take account of actual volumes discharged and qualitative factors (such as concentration of toxic substances), and discriminate against the most harmful emissions. In some cases, however, discriminating approaches are not practical and more blunt instruments (such as taxes on inputs or outputs) must be used.

Matching instruments and implementation capacity. New instruments should be developed within the context of existing regulatory and institutional frameworks. Their type and scope must match the institutional capacity to implement them. Pollution charges can, for example, be designed to be collected and managed by environmental protection agencies, which are responsible for monitoring and enforcement of existing laws are good mechanisms. Pollution taxes "built in" to the national fiscal system and collected by fiscal authorities may be more complex to implement. China, for example, first experimented with the first approach for several years before moving to the second.

Allocating the proceeds. Principles of public finance argue against allocated revenues from taxes to pre-determined purposes (earmarking). At the same time, allocating at least a share of the proceeds from environmental taxes to monitoring and enforcement activities (partial earmarking) can be justified. In such cases, it is important to avoid over-reliance on such revenues, which may come to reduce the incentives for the agency to reduce pollution levels, in order to preserve its financial resource base. Allocating some part of the proceeds from pollution taxes to the industry ("revenue recycling") may also be considered, in order to make the introduction of pollution charges politically more acceptable. In such cases, it is essential for the financial reflows to industry to discriminate between good and bad performers, and discourage continued bad practice. Revenues may be used to encourage investment in environmentally preferable equipment and methods.

Setting the level of taxes. Ideally, from an environmental standpoint, a tax should be equal to marginal cost of abatement to provide incentives for actual pollution reduction. If it is lower than that, firms will generally prefer to pay the tax. The charges would thus operate primarily as revenue-raising mechanism with little pollution-reduction benefits. In practice, however, low levels of taxes or charges can help establish the principle that industries should pay for pollution and allow experimentation with new instruments. Experience from China and eastern Europe suggests that even low levels of taxes can have important incentive impacts, paving the way for progressively higher charges (see Box 10.2). To maximise actual impact while minimising administrative burden, it is generally better to set significant taxes/charges on a narrow selection of pollutants/effluents, concentrating on the most hazardous, rather than low levels of taxes/charges on a large number. Indexing the taxes to inflation is also necessary to avoid them losing their relevance over time.

Interests and perspectives of key stakeholders

Interests of the poor. The urban poor suffer disproportionately from air and water pollution in the form of *e.g.* respiratory and water-borne diseases. They often live in the most contaminated areas, including in the vicinity of waste disposal sites, or work in the most hazardous industries. At the same time, those employed in those industries may perceive restrictions placed on their employers as threats to their jobs. Thus industrial pollution control can be a divisive issue amongst different low-income groups, depending on their occupation and degree of direct exposure.

Non-poor households. Although they are not the most exposed, relatively better off groups are often more aware of the health impact of pollution. In many countries, middle-class groups have played leading roles in pressuring governments to address the issue of industrial pollution.

Large-scale industries. Industries will generally resist anything that adds to their costs of production. For large-scale industries, and in particular multinational enterprises, pollution charges and other costs associated with environmental management account for a small share of total costs and can readily be absorbed. Evidence from OECD country-industries suggests that pollution charges do not impose significant costs to industries, and are not a significant factor in their decisions of where to locate. In the medium to long run, cleaner production approaches, which reduces pollution by using inputs more efficiently may actually reduce costs. However, industries will often express the concern that pollution charges will reduce their competitiveness relative to foreign competitors who do not have to pay them. Faced with a choice between pollution taxes and quantitative limits, industries generally prefer taxes which give them flexibility. In certain circumstances, such as in peak production periods, they may prefer to pay the tax rather than to reduce pollution to certain fixed levels. This flexibility is critical for plant operators.

Medium and small-scale industries. Smaller firms generally face more difficulties in adjusting to pollution charges and other forms of environmentally-related constraints. They generally have less access to the necessary know-how to adapt to cleaner production methods. In highly competitive or declining industries, where profit margins are low, pollution charges can represent a significant additional burden, notably for the poorest performers, which often operate very inefficiently and with obsolete equipment, generating the most pollution. In such cases, compensatory measures may be needed to facilitate adjustment. In the case of export dependent industries, adjustment to more rigorous environmental

Box 10.1. **Environmental standards in importing countries help improve the Indian leather industry**

Leather tanning can be highly pollution-intensive, due to the use of such inputs as formaldehyde, cadmium, chromium and pentachlorophenol, which are carcinogenic and can end up in the final product as residues. The adoption by several OECD countries of residue limits for chemicals used in the manufacture of leather goods has put increasing pressure on the leather industries in countries such as Argentina, Brazil, China, India and Pakistan.

For example, when Germany introduced its prohibition on pentachlorophenol (PCP) in 1989, the Indian leather industry was caught unaware. But for exporters it was imperative to adjust. The government intervened and banned manufacture and use of PCP and manufacturers turned to alternative inputs. The Government of India, UN agencies and bilateral donors, including Germany, also provided considerable technical and financial assistance. The environmental performance of the industry improved, notably with respect to air and water pollution. According to India's Council of Leather Exporters, most tanneries in Tamil Nadu, the leather capital of India, now have access to an effluent treatment plant. Occupational safety and health standards also improved. Thus, the standards adopted by industrialised countries actually helped to speed up improvements in pollution control and overcome industry's resistance to change.

Source: "Limits on chemical residues in leather goods". Documentation for the OECD Global Forum on Trade, Workshop on Environmental Requirements and Market Access: Addressing Developing Country Concerns, New Delhi, 27-28 November 2002.

standards may be a precondition to continued access to developed countries markets (see Box 10.1).

Agencies responsible for water management. Water supply agencies have a direct interest in preventing pollution of water bodies which they rely on but have no control over. Municipalities often provide wastewater and other treatment services. The imposition of cost-recovery charges, at least to cover running costs, would free up funds for other uses.

Interest of the Administration. *Finance ministries* have a strong interest in environmental taxes which are allocated to the budget rather than earmarked for environmental purposes. They will generally favour taxes which are administered by fiscal authorities rather than environmental agencies. *Environmental agencies* have the main responsibilities for monitoring factories' environmental performance and enforcing regulations. Sometimes, they are averse to market based instrument, preferring familiar regulatory approaches. This attitude can change significantly if the proceeds from such instruments are partially earmarked to fund monitoring and enforcement activities. The bargaining power of local environmental agencies relative to the firms they monitor is stronger when the social impact of a firms' emissions are high and concentrated on a given area or community, and relatively lower when those industries or firms face serious financial difficulties.

Citizens groups and NGOs. Civil society groups are often at the forefront of initiatives to control pollution. Women's groups, in particular, have often been instrumental in raising awareness of the health impacts, notably on children, and have been powerful agents for change. In general, the ability of citizens to litigate to recover damages from pollution can be a significant financial and public-relations incentive for firms to avoid causing those damages.

Donors. In the case of transboundary pollution (such as *e.g.* when sulphur emissions from China are linked to acid rain in Japan and Korea), donors may have an immediate interest in helping developing countries partners tackle the issue of pollution control, including through technology co-operation to accelerate the dissemination of cleaner production processes.

Managing the reform process: key elements

Public awareness and participation. Public pressure invariably plays a key role in building the necessary political will to enact and enforce pollution regulation, and overcome resistance of industry. A key requirement is the public disclosure of information on the health hazards from pollution as well as on industry's impacts on the environment. Such information must be made available to the public in ways that are easy to understand.

Helping firms comply with standards. There are many low-cost ways for industrial producers to reduce pollution and waste. Compared with standard methods, "cleaner production techniques" use energy, raw materials and other inputs more efficiently, leading to cost savings. By focussing on prevention, cleaner production approaches reduce the need for "end of pipe" pollution control equipment. Efforts to disseminate these approaches focus on increasing awareness amongst firms of their financial benefits, notably by working through industry associations.

"Shaming polluters". In some countries, publishing information on the performance of individual firms is an important way to pressure them. In Indonesia, a system of public rating for polluters has significantly improved performance and attracted strong popular support. Similar approaches are working in the Philippines and in Columbia (World Bank, 2000).

Capacity and Credibility of Environmental Agency. Focussing on few highly critical issues where visible success can be achieved, is essential for enforcement agencies to minimise risk of failure, and to build support for future steps. This generally implies focussing on a limited number of sources (*e.g.* a few large factories), pollutants (*e.g.* heavy metals) and medias (*e.g.* rivers).

Monitoring agencies must also have accurate and timely information on industrial pollution flows, their origins and their impacts on, *e.g.* water and air quality. Reports on effluents submitted by firms must be carefully reviewed for accuracy, since the scope for false or under-reporting and corruption is high. Agencies must also be able to face legal challenges from industries, for example in reaction to the disclosure of information about their effluents, by providing solid data. Partial earmarking of pollution taxes can help develop and maintain the capacity of the environmental monitoring agency and protect it from politically motivated attacks on its budget.

Sequencing reforms and combining instruments: Scheduling and announcing future increases of charges or taxes in advance and allocating the proceeds to help firms invest in pollution control can facilitate adjustment and reduce resistance to the introduction of charges. Gradual approaches also provide room for the necessary consultation with affected industries. In China, a combination of pollution taxes and subsidies for pollution abatement, designed and introduced gradually, is proving its effectiveness (Wang Hua, 1999).

Box 10.2. **Pollution levies in China and in the countries of Central and Eastern Europe (CEES)**

China's pollution levy system (PLS) is among the most extensive in the world. It is an example of pragmatic and gradual implementation of EFR, in the context of a transition towards a market-based economy. The scheme began in 1979. Initially confined to only a few provinces, it has expanded over time, building on the lessons from implementation experience. By 1994, over USD 2 Billion had been collected from environmental levies. The system has been regularly monitored and amended in light of weaknesses identified, with respect to the level of the levies, enforcement difficulties and others as well as the tradeoffs faced by EPBS between reducing emissions and generating revenue. The PLS does not conform to a "textbook" example of environmental taxation. For example, fees are paid only for discharges exceeding a certain level, thus resembling non-compliance fees. In addition, the funds collected are used first to finance abatements expenditures by industry and for central administrative costs. While the fees are considered to be lower than marginal abatement costs, effectiveness of collection is linked to population density and income levels, suggesting that public pressure plays an important role in stimulating enforcement efforts. Despite uneven progress in different parts of the country, the system is generally considered to play an important role in containing pollution in China in a period of rapid industrialisation.

By the time they began the transition to market economies, **CEE countries** had built up significant environmental hazards (including, for example, toxic waste sites). These countries used Environmental Funds as the main mechanism to mobilise the necessary resources for cleaning up. Although donors such as the GEF, World Bank and bilateral agencies provided considerable assistance, including by facilitating debt-nature swaps, these funds have relied heavily on environmental taxes and fees. From the strictly environmental point of view, these taxes were initially set far below optimal levels. From a political and social point of view, it would not have been feasible to set taxes commensurate with the very high environmental damages. The risk would have been to many firms to insolvency. These taxes played important revenue raising roles, helping finance clean-up costs and also helped foster pollution prevention and abatement.

Source: Sterner, 2003.

References

Abed et al. (1998), "Fiscal Reform in Low-Income Countries: Experience under IMF Supported Programs", *IMF Occasional Paper* No. 160.

van Beers, C. and A. de Moor (2001), *Public Subsidies and Policy Failures*. Cheltenham: Edward Elgar.

Beslier S. (2004), "Enforcement and surveillance: What are our technical capacities and how much are willing to pay?" In *Fish Piracy, Combating Illegal, Unreported and Unregulated Fishing*, OECD, Paris, 2004.

Bojo, J. and R.C. Reddy (2003), *Poverty Reduction Strategies and Environment – A Review of 40 Interim and Full Poverty Reduction Strategy Papers*.

Briscoe, J. (1999), "The Changing Face of Water Infrastructure Financing in Developing Countries", *International Journal of Water Resources Management*, 15:3, pp. 301-308.

Datt, Garg and Narang (2003), *Environmental fiscal reform in India: Scope, Challenges and Opportunities*, paper presented at OECD Scoping Workshop, Paris, January 2003.

DFID (Department for International Development) (2000), *Eliminating world poverty: making globalisation work for the poor*, DFID, UK.

DFID, EC (European Commission), UNDP (United Nations Development Programme) and WB (World Bank) (2002), *Linking Poverty reduction and environmental management: policy challenges and opportunities*, DFID, UK.

Dinar, A. (2000), *Political Economy of Water Pricing Policies*. World Bank.

Dinar, A. and A. Subramanian (1997), *Water Pricing Experiences. An International Perspective*. World Bank Technical Paper No. 386.

EC (2002), Green Paper, "The Common Fisheries Policy after 2002".

Essama-Nssah B. and J.J. Gockowski (2000), *Cameroon, Forestry Sector Development in a difficult political environment*, OED (The World Bank Operations Evaluation Department).

FAO (Food and Agriculture Organisation of the United Nations) (1983), *Report of the Expert Consultation on the Regulation of Fishing Effort*, Rome, 17-26 January 1983, FAO Fisheries Report No. 289.

FAO (2002), *The State of World Fisheries and Aquaculture 2000*. FAO, Rome.

Gatto, F.D., M. Richards and G.A. Lopez (2003), *The cost of illegal logging in Central America. How much are the Honduran and Nicaraguan Governments losing?*

Gautam M. et al. (2000), *Indonesia: The challenges of World Bank Involvement in forests*, The World Bank Operations Evaluation Department (OED), *www.worldbank.org/html/oed*.

Global Witness (2002), *Forest Law enforcement in Cameroon*.

Government of Georgia (2000), "Poverty Reduction and Economic Growth Program of Georgia – Intermediary Document", *http://poverty.worldbank.org/files/georgia%20IPRSP2.pdf*.

Gray, J.A. (2002), "Forest Concession Policies and Revenue Systems – Country Experience and Policy Changes for Sustainable Tropical Forestry", World Bank Technical Paper No. 522.

GTZ (Deutsche Gesellschaft für Technische Zusammenarbeit – Germany Agency for Technical Co-operation) (1999), "Internalisierung externer Kosten im Energie- und Transportsektor", Erfahrungen in Industrieländern und deren Anwendung auf Entwicklungsländer, Internes Diskussionspapier und Materialiensammlung, Eschborn.

GTZ (2001), Fuel Prices and Vehicle Taxation.

Gutman, Pablo (ed.) (2003), From Goodwill to Payments for Environmental Services. A Survey of Financing Options for Sustainable Natural Resource Management in Developing Countries, a joint project of DANIDA/WWF.

IEA (International Energy Agency) (1995), Middle East Oil and Gas. IEA, Paris.

IEA (1999), "Looking at Energy Subsidies: Getting the Prices Right", World Energy Outlook, IEA, Paris.

IEA (2004), World Energy Outlook, IEA, Paris.

IEA/UNEP (United Nations Environment Programme) (2002), Reforming Energy Subsidies.

IEEP (Institute of European Environmental Policy) (2003), Fisheries agreements with third countries – is the EU moving towards sustainable development?, IEEP.

IFREMER (Institut français de recherche pour l'exploitation de la mer) (1999), Evaluation of Fisheries Agreements Concluded by the European Community. Summary report.

IMF (International Monetary Fund) (2000), "The IMF and Environmental issues", Issues brief. Also available at: www.imf.org/external/np/exr/ib/2000/041300.htm.

IMF (2001), "IMF Team Keeps an Eye on Linkages Between Environment and Macroeconomic Policies", IMF Survey, Vol. 30, No. 21, IMF, Washington. Also available at: www.imf.org/external/pubs/ft/survey/survey01.htm.

IMF, UNEP and WB paper (2002), Financing for Sustainable Development.

Islamic Republic of Mauritania (2000), Mauritania Poverty Reduction Strategy Paper: http://poverty.worldbank.org/files/mauritania_prsp.pdf.

Karsenty A. et al.(2000), Economic and financial audit of the forestry sector in Cameroon (in French, summary in English), DFID, UK.

Karsenty, A. (2000), "Economic Instruments for Tropical Forests – The Congo Basin" Centre for International Forestry Research.

Leruth, L., R. Paris and I. Ruzicka (2001), "The Complier Pays Principle: The Limits of Fiscal Approaches Toward Sustainable Forest Management" IMF Staff Papers, Vol. 48, No. 2.

Ma, Zhong (2003), "Environmental fiscal reform in China", paper presented at OECD Scoping Workshop, Paris, January 2003.

Metschies, G. (2001), Fuel Prices and Vehicle Taxation, p. 64.

Metschies, G. (2003), Presentation at Annual Stakeholders Meeting of Sub-Saharan Africa Transport Policy Program (SSATP) in Kigali, Rwanda, 25-30 May 2003.

Moor, A.P.G. de and P. Calamai (1997), Subsidising Unsustainable Development: Undermining the Earth with Public Funds. Earth Council.

Morden, C. (2003), "Environmental fiscal reform: Options for South Africa", presented at OECD Scoping Workshop, Paris, January 2003.

Nichol, P. (2003), "A developing country puts a halt to foreign overfishing". Economic Perspectives, An Electronic Journal of the US Department of State, Vol. 8, No. 1.

O'Halloran, E. and V. Ferrer (1998), *The evolution of Cameroon's new forestry legal, regulatory and taxation system, Manuscript,* World Bank, January 1998.

OECD (1997), *Environmental Taxes and Green Tax Reform.* OECD, Paris.

OECD (1999), *Environmental Taxes, Recent Developments in China and OECD Countries,* OECD, Paris.

OECD (2000), *Shaping the Urban Environment in the 21st Century*: A DAC Reference Manual on Urban Environmental Policy. Available from *www1.oecd.org/dac/urbenv/,* OECD, Paris.

OECD (2001), "Poverty-Gender-Environment Linkages", *DAC Journal 2001,* Vol. 2 No. 4, OECD, Paris.

OECD (2001a), *Environmental related taxes; issues and strategies,* OECD, Paris

OECD (2001b), *Sustainable Development – Critical Issues,* OECD, Paris.

OECD (2001c), *An Overview of Green Tax Reform and Environmentally Related Taxes in OECD Countries,* OECD, Paris.

OECD/DAC (2002a), "Development Co-operation Review of Spain", *DAC Journal 2002,* Vol. 3, No. 2, OECD, Paris.

OECD (2002b), "Subsidies and the Environment: An Overview of the State of Knowledge", *Joint Working Party on Trade and Environment,* OECD, Paris.

OECD (2003b), Good Practices of Public Environmental Expenditure Management in Transition Economies. OECD, Paris.

OECD (2005), "Environmentally Harmful Subsidies: Analysis and Assessment" (forthcoming), OECD, Paris.

Porter, G. (2002), *Fisheries subsidies and over-fishing: towards a structured discussion,* UNEP.

van Santen, Gert (2001), *Governance of marine fisheries and its impact on rural poverty – past, present and future.*

Sayeg, P. (1998), "Successful Conversion to Unleaded Gasoline in Thailand" *World Bank Technical Paper* No. 410.

Seymour, F., and N. Dubash (2000), *Right conditions: The World Bank, structural adjustment, and forest policy reform.* World Resource Institute, Washington.

Sterner, T. (2003), Resources for the Future Press. *Policy Instruments for Environmental and Natural Resource Management.*

UN (United Nations) (2002), *Johannesburg Plan of Implementation.*

UNEP (2003), *Energy Subsidies: Lessons Learned in Assessing their Impact and Designing Policy Reforms.*

UNEP (2004), Opportunities and Challenges for the Use of Economic Instruments in Environmental Policy.

Wang, Hua (1999), *How the Chinese System of Charges and Subsidies Affects Pollution Control Efforts by China's Top Industrial Polluters.* World Bank Working Paper 2198.

World Bank (1994), *World Development Report 1994 – Infrastructure for Development,* Oxford University Press, New York.

World Bank (1996), Cambodia, Forestry Policy Assessment, Report No. 5777-KH.

World Bank (2000), *Greening Industry, New Roles for Communities, Markets and Governments.*

World Bank (2001), *A User's Guide to Poverty and Social Impact Analysis.* World Bank Poverty Reduction Group (PRMPR) and Social Development Department (SDV).

World Bank (2002), *World Development Report 2002*.

World Bank (2003), *World Development Report 2003*.

WRI (World Resource Institute) (2002), Power Politics, Equity and Environment in Electricity Reform.

ENVIRONMENTAL FISCAL REFORM FOR POVERTY REDUCTION – ISBN 92-64-00868-3 – © OECD 2005

OECD PUBLICATIONS, 2, rue André-Pascal, 75775 PARIS CEDEX 16
PRINTED IN FRANCE
(43 2005 16 1 P) ISBN 92-64-00868-3 – No. 54031 2005